Atmospheric Air Pollution and Monitoring

Edited by Abderrahim Lakhouit

Published in London, United Kingdom

IntechOpen

Supporting open minds since 2005

Atmospheric Air Pollution and Monitoring
http://dx.doi.org/10.5772/intechopen.77883
Edited by Abderrahim Lakhouit

Contributors
Tabbsum Hanif Mujawar, Lalasaheb Deshmukh, Steven Cryer, Ian Van Wesenbeeck, Yahaya Abbas Aliyu,
Joel Botai, Aliyu Abubakar, Terwase Youngu, Jimoh Suleiman, Mohammed Shebe, Muhammed Bichi, Ivan
Nedkov, Abderrahim A.L Lakhouit, Douha Belaidi, Hanaâ Hachimi, Aouatif Amine

Notice
Statements and opinions expressed in the chapters are these of the individual contributors and not
necessarily those of the editors or publisher. No responsibility is accepted for the accuracy of
information contained in the published chapters. The publisher assumes no responsibility for any
damage or injury to persons or property arising out of the use of any materials, instructions, methods
or ideas contained in the book.

First published in London, United Kingdom, 2020 by IntechOpen
IntechOpen is the global imprint of INTECHOPEN LIMITED, registered in England and Wales,
registration number: 11086078, 7th floor, 10 Lower Thames Street, London,
EC3R 6AF, United Kingdom
Printed in Croatia

British Library Cataloguing-in-Publication Data
A catalogue record for this book is available from the British Library

Additional hard and PDF copies can be obtained from orders@intechopen.com

Atmospheric Air Pollution and Monitoring
Edited by Abderrahim Lakhouit
p. cm.
Print ISBN 978-1-78985-279-0
Online ISBN 978-1-78985-280-6
eBook (PDF) ISBN 978-1-78984-172-5

We are IntechOpen,
the world's leading publisher of
Open Access books
Built by scientists, for scientists

4,700+
Open access books available

121,000+
International authors and editors

135M+
Downloads

Our authors are among the

151
Countries delivered to

Top 1%
most cited scientists

12.2%
Contributors from top 500 universities

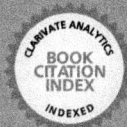

Interested in publishing with us?
Contact book.department@intechopen.com

Numbers displayed above are based on latest data collected.
For more information visit www.intechopen.com

Meet the editor

Dr. Abderrahim Lakhouit has a PhD in Civil and Environmental Engineering from the University of Sherbrooke, Quebec, Canada. He has two master's degrees in Environmental Engineering and Renewable Energy and Energy Efficiency. He is an assistant professor at the University of Tabuk, Kingdom of Saudi Arabia. Previously he worked as a teaching assistant at Canadian universities. Dr. Lakhouit is also a researcher and has published articles in international journals such as Chemosphere. He is an associate and guest editor as well as reviewer for many international journals, including Waste Management, Environments, and others. He is an active member in the Association of Environmental Engineering and Science Professors (AEESP).

Contents

Preface

Quality of life and air quality are inextricably linked. Air pollution is a growing concern worldwide as exposure to air pollution can be harmful for human health. Indoor air quality (IAQ) in particular is a major environmental concern. In fact, people spend approximatively 90 percent of their time indoors where they are exposed to chemical and biological contaminants and possibly carcinogens. There are many virus and microbes that we can eliminate or reduce by using ventilation.

This book is a study of atmospheric air pollution and presents ways we can reduce its impacts on human health. It discusses tools for measuring IAQ as well as analyzes IAQ in closed buildings. It is an important documentation of air quality and its impact on human health.

Dr. Abderrahim Lakhouit
University of Tabuk,
Saudi Arabia

Chapter 1

Introductory Chapter: Indoor Air Quality in the Closed Building

Abderrahim Lakhouit

1. Introduction

Air pollution is a major concern that has been recognized throughout the world during the nineteenth and twentieth century and until now. In the Middle Ages, the heavy industry and the burning of coal in cities, especially in Europe (Belgium, French, Germany, Poland and the United Kingdom), released into atmosphere numerous chemical compounds as carbon dioxide and sulphur dioxide. In the late eighteenth century, the Industrial Revolution, beginning in the United Kingdom (UK), led to escalation in pollutant emissions based around the use of coal by both homes and industry. The sky of certain cities in Europe was covered by smoke. The air pollution in this period in UK was given the dramatic episodes known in the literature by 'London Smog'. Air quality is influenced by a variety of factors and is a complex issue according to many authors in this field. Air pollutants can be represented by a numerous of substances present in the atmosphere at concentrations above their normal background levels which can have a measurable effect on humans, animals and vegetation. The main aim of the Air Quality Index (AQI) is to help deciders or governments to understand what local air quality means to human health. The governments and non-government agencies develop air emission standards for many air pollutants. The main objective of emission standards is to establish quantitative limits on the permissible concentrations of specific air pollutants. The standards are determined in order to protect human health and environment. To determine the standard for each pollutants, many factors should be taken in account. For example, the toxicity of pollutant should be considered. More attention should be given to how the pollutant enters to the organism.

Air pollution can be defined as: the introduction of particulates (particulate matter or solid particles), biological molecules or other harmful chemical or materials into our atmosphere [1]. The air pollution can cause diseases, death to humans and damage to the environment and ecosystems. The source of air pollution can be from anthropogenic or natural Sources. The air quality can be measured by indexes. In the literature, two types (outdoor and indoor) of air pollution can be found. The present chapter is focused on indoor air quality. The indoor air quality (IAQ) is a very important aspect in the design of buildings due to the dominant exposure for humans.

2. Air quality in hospital

The air quality in the medical building is treated in this chapter. This is an important point that was raised in this chapter. Until recently, the health effects of indoor air pollution have received relatively little attention. In particular, air quality at hospitals is probably a risk factor with serious health consequence on the working staff, patients and visitors. Infection is a common event in hospitals, and many studies have investigated the levels, sources and characteristics of bioaerosol in hospital [2, 3]. So the hospital sector is not shielded from the problems bound to inside air quality. Due to multiple sources of pollution and the presence of vulnerable people, this sector is particularly at risk. Moreover, outdoor air is an important source of pollution, which affects the IAQ [4].

More than 2 million people in Europe are infected due to health care-associated infection (HAI) [5]. Poor hospital IAQ may cause headaches, fatigue, eye, and skin irritations and other symptoms. Although it is believed that transfer of infection by direct contact is the main cause for HAI, there are evidences that airborne bacteria may also cause infection due to inhalation of such bacteria. Therefore, it is essential to understand the dynamics of infectious particles due to respiratory diseases such as severe acute respiratory syndrome (SARS) and tuberculosis (TB).

3. Air quality in the closed buildings

In emergent countries, rapid growth in the global population requires expansion of building stock. This demand varies in time and also between different buildings; yet, conventional methods are only able to provide the comfort thermal in the building. To save energy in the building is an additional challenge. To save energy and to have a good air quality in the building are the big deals especially in cold countries. In recent years, the construction of tighter building envelopes, the increase in office equipment and the widespread use of synthetic materials have ensured that the concerns about indoor air quality are increasing. The sick building syndrome (SBS) is used, for the first time, to explain a situation in which the occupants of a building experience acute health- or comfort-related effects that seem to be linked directly to the time spent in the building. This issue is related to air quality in the building. The building is developed to save energy. The air circulation in this type of building is not good. The air spent many time in the building. The term of age of air was used for the first time to describe the time spent by the mass of air in the building. The complainants may be localized in a particular room or zone or may be widespread throughout the building. Generally, people spend about 90% of their time inside buildings [6]. To ensure the air quality in the building, Scientifics suggest using mechanical ventilation. The strategy of ventilation and its efficiency should be studied more in order to have excellent air quality with minimum of energy.

Author details

Abderrahim Lakhouit
Civil Engineering Department, Faculty of Engineering, University of Tabuk, Tabuk, Saudi Arabia

*Address all correspondence to: abderrahim.lakhouit@usherbrooke.ca

IntechOpen

References

[1] Dominici F, Peng RD, Bell ML, et al. Fine particulate air pollution and hospital admission for cardiovascular and respiratory diseases. Jama. 2006;**295**(10):1127-1134

[2] Hoseinzadeh E, Samarghandie MR, Ghiasian SA, et al. Evaluation of bioaerosols in five educational hospitals wards air in Hamedan, During 2011-2012. Jundishapur Journal of Microbiology. 2013;**6**(6)

[3] Li C-S, Hou P-A. Bioaerosol characteristics in hospital clean rooms. Science of the Total Environment. 2003;**305**(1-3):169-176

[4] Jung C-C, Wu P-C, Tseng C-H, Su H-J. Indoor air quality varies with ventilation types and working areas in hospitals. Building and Environment. 2015;**85**:190-195

[5] Pittet D, Allegranzi B, Sax H, Bertinato L, Concia E, Cookson B, et al. Considerations for a WHO European strategy on health-care-associated infection, surveillance, and control. The Lancet Infectious Diseases. 2005;**5**(4):242-250

[6] Lakhouit A. Modelisation de la Qualite de l'air Dans une Unite de Bronchoscopie: Influence des Strategies de Ventilation. Canada: Ecole de Technologie Superieure (Canada); 2011

Numerical Analysis of Indoor Air Quality in Hospital Case Study: Bronchoscopy Unit

Hanaâ Hachimi, Chakib El Mokhi, Badr T. Alsulami and Abderrahim Lakhouit

Abstract

This paper presents three ventilation scenarios for a bronchoscopy unit using a numerical study. A Fire Dynamics Simulator (FDS) is employed for this purpose. The results obtained are visualized using Smokeview (SMV), which is a program for displaying FDS results. The numerical results are compared with experimental ones from Cheong and Phua's research study. This study was chosen because it investigates ventilation strategies in hospital isolation rooms using a tracer gas technique. In the present work, six points of measurements are utilized to evaluate the concentrations of contaminants and air velocity. The results show that the concentrations estimated by FDS are inferior to the experimental results given by Cheong and Phua . For example, in the SP1 point of measurement, the concentrations estimated by FDS and by Cheong and Phua are 20 and 28.9 ppm, respectively, while in the SP5 point, the concentrations estimated by FDS and by Cheong and Phua are 28.6 and 32.9 ppm, respectively. The error percentages between FDS estimates and experimental measurements made by Cheong and Phua range between 1 and 32%.

Keywords: modelling, indoor air quality, ventilation

1. Introduction

Indoor air quality (IAQ) is an important aspect in the design of buildings due to the effect of IAQ on human health and well-being [1, 2]. In the developed world, people spend about 90% of their time indoors [3], where they are exposed to chemical and biological contaminants and possibly also to carcinogens [4]. Until recently, the health effects of indoor air pollution have received relatively little attention [5].

In particular, the air quality at hospitals carries with it a risk factor for serious health consequences not only for the medical staff but also for patients and visitors. Infection is a common event in hospitals, and many studies have investigated the levels, sources, and characteristics of bioaerosols in these settings [6]. Due to multiple sources of pollution and the presence of vulnerable people, the health sector is particularly at risk of low IAQ [7]. Additionally, outdoor air pollution can affect indoor air quality [8]. Since hospitals are primarily and traditionally a place for people to recover from illness or disease, improving IAQ can help reduce recovery times and thus boost overall productivity [9].

In Europe, more than two million people annually become infected due to healthcare-associated infection (HAI) [10]. Although direct contact is believed to be the main route for transfer of HAI, there is evidence that airborne bacteria may also cause infection due to inhalation [11]. Therefore, it is essential to understand the dynamics of infectious particles that are present in respiratory diseases such as severe acute respiratory syndrome (SARS) and tuberculosis (TB).

This paper examines infection transfer in hospitals caused by poor air quality. To date, only a few researchers have investigated this topic. Cheong and Phua [1] conducted an experimental study on ventilation strategies for improving indoor air quality inside a hospital isolation room, and Qian and Li [12] analyzed ventilation strategies in a hospital isolation chamber using numerical digital tools.

The purpose of the present study is to develop a model for air velocity, temperature profile, and some concentrations of gaseous contaminants in a hospital bronchoscopy unit. Modeling is done using large-eddy simulation (LES). The outcomes of this study will be applied to a hospital in Morocco (North Africa). To validate the work, a hospital in Kenitra, Morocco, has been chosen to measure air quality in the bronchoscopy unit.

2. Materials and methods

To investigate the indoor air quality in the bronchoscopy unit, a Fire Dynamics Simulator was used to simulate the movement of air and the temperature profiles.

2.1 Fire Dynamics Simulator (FDS)

An FDS is a computational fluid dynamics (CFD) model of fire-driven fluid flow. The software described in this document numerically solves a form of the Navier–Stokes equations appropriate for low-air velocity thermally driven flow, with an emphasis on smoke and heat transport from fires.

In this study, the FDS software (version 5) was used. FDS is developed by the National Institute of Standards and Technology (NIST). The first version of FDS was publicly released in February 2000. To date, about half of the applications of the model have been for the design of smoke handling systems and sprinkler/detector activation studies. The other half consists of residential and industrial fire reconstructions [13].

2.2 Hydrodynamic model

The Hydrodynamic Model FDS numerically solves a form of the Navier–Stokes equations, which are appropriate for low-air velocity thermally driven flow, with an emphasis on smoke and heat coming from fires. The core algorithm is an explicit predictor–corrector scheme with second-order accuracy in space and time. Turbulence is treated by means of the Joseph Smagorinsky form of Large-Eddy Simulation. The LES is the default mode of operation [14].

The equations of conservation of mass and momentum written in a system of Cartesian coordinates are as follows [15]:

Equation of conservation of mass:

$$\frac{\partial \rho}{\partial t} + u.\nabla \rho = -\rho \nabla .u \tag{1}$$

where ρ = air density (kg/m^3); u = flow velocity; ∇ = operator nabla.

Equation of conservation of momentum:

$$\frac{\partial u}{\partial t} + u \times \omega + \nabla \mathcal{H} = \frac{1}{\rho}\left((\rho - \rho_0)\vec{g} + f_b + \nabla \cdot \tau_{ij}\right) \tag{2}$$

where $\mathcal{H}$ = pressure; $\vec{g}$ = vector of gravity (m/s²).
Such that:

$$(u.\nabla)u = \frac{\nabla|u|^2}{2} - u \times \omega \tag{3}$$

ω = air velocity component on z (m/s).

$$\mathcal{H} = |u|^2/2 + p/\rho_\infty \tag{4}$$

ρ_∞ = density of ambient air (kg/m³)

$$\delta_{ij} = \begin{cases} 1 \text{ si } i = j \\ 0 \text{ si } i \neq j \end{cases} \tag{5}$$

δ_{ij} = Kronecker delta

$$S_{ij} = \frac{1}{2}\left(\frac{\partial u_i}{\partial x_j} + \frac{\partial u_j}{\partial x_i}\right) \tag{6}$$

S_{ij} = strain tensor symmetric rate.

2.3 Smokeview

Smokeview is a software developed as FDS by NIST. It is used to view the geometry, mesh size, and results obtained by FDS and includes several visualization techniques.

2.4 Geometry of bronchoscopy unit

The studied room, as shown in **Figure 1**, consists of a chamber and a corridor. The chamber measures 4.80 m in length, 3.35 m in width, and 2.5 m in height. It

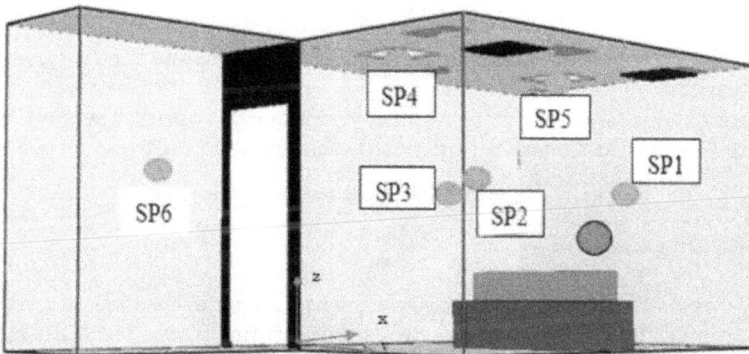

Figure 1.
Geometry of bronchoscopy.

Positions		SP1	SP2	SP3	SP4	SP5	SP6
Coordinates (m)	x	2.30	1.05	2.30	1.50	1.5	0.60
	y	3.90	2.40	1.03	3.90	1.3	−1.32
	z	1.40	1.40	1.40	2.50	2.5	1.40

Table 1.
Position of measurement points in bronchoscopy unit.

contains a bed in the middle. The corridor is 2.65 m long and 1.2 m wide. The corridor is the same height as the chamber.

The room is classified as a high-risk setting (Class 1). According to international standards and the American Society of Heating, Refrigerating and Air Conditioning Engineers (ASHRAE) standards ([16], p. 62), the room must be maintained under negative pressure. This is to prevent any exfiltration of contagious microorganisms or antibiotic-resistant bacteria that may be emitted by a sick patient to other parts of the hospital, as these organisms and bacteria might infect other patients, medical staff, or even visitors.

We use here six points of measure: SP1, SP2, SP3, SP4, SP5, and SP6. These points fall in the plane z = 1.4 m (**Table 1**).

2.5 Mesh

For a typical building design simulation using FDS, a large volume of space is simulated. In describing this computation volume, one or more subsections of the overall volume are referred to as a "mesh" and entered as "&MESH" in the FDS input file. In many cases, multiple meshes of different resolutions are required to accurately define the simulated domain. Most modern computers have multiple "cores" or processors per CPU chip, and FDS through the Message Passing Interface (MPI) feature allows for each mesh to be assigned to a specific processor using the MPI_PROCESS keyword. This feature enables any number of meshes to be assigned to the same processor to improve simulation efficiency. In the present study, three different meshes have been tested:

1. Dense mesh (DM)

2. Less dense mesh (LDM)

3. Coarse mesh (CM)

Note that DM has 2.4 times more mesh nodes than LDM and 7.3 times more nodes than CM.

Figure 2 shows LDM according to the x-y, x-z, and y-z planes. The mesh is subdivided in two parts: mesh for the closed room (zone 1) and mesh for the corridor (zone 2).

2.6 Ventilation scenarios

Three ventilation scenarios are used. In the first scenario, the isolated room is supplied with fresh air by two square diffusers measuring 0.6 m. The diffusers are at the sides of the room and are located in the ceiling. The air is extracted from the room by two square grilles also measuring 0.6 m and also positioned at the sides of the room and located in the ceiling. The objective of this strategy is to dilute the

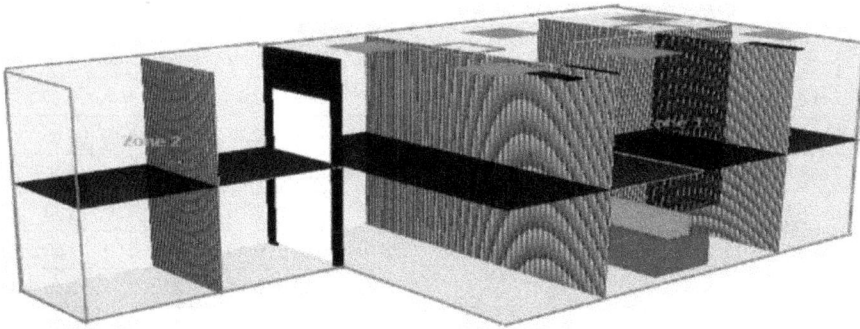

Figure 2.
LDM mesh.

contaminant as effectively as possible in order to obtain a uniform concentration of sulfur hexafluoride (SF6) across the entire volume of the room. In the first scenario, the contaminant is released in the room at a flow rate of 0.31 L/min. Numerically, we impose the release of the contaminant evenly on the six sides of a cube of 27 cm^3 as a condition. We obtain the flow volume by a surface unit of 0.97×10^{-3} m^3. Cheong and Phua [1] present some numerical and experimental results for the first scenario, which we demarcate as "scenario 1."

In "scenario 2" the diffusers' positions are similar to those in scenario 1. However, the air is extracted by two mural grilles located 30 cm above the floor near the patient's bed. In this scenario, SF6 is released at a rate of 0.63 l/min (1.94×10^{-3} m^3/ s/m^2). This second ventilation strategy aims at creating a flow from the top of the room to the floor. The aim is to quickly direct the gaseous contaminant "rejected" by the patient towards the extraction grilles. Numerically, it is very difficult to simulate the structure to the output of a diffuser of flow.

"Scenario 3" is similar to scenario 2, except that the supplying diffusers are replaced by blow grilles located on the ceiling directly above the patient. Unlike the diffusers, the air in scenario 3 is blown directly to the floor.

The component positions of the modeled room are presented in the table below.

3. Results and discussion

3.1 Scenario 1: Results and discussion

In the first ventilation scenario, two different software (FDS and Smokeview) allow the exploitation of a quantity of information at the end of every simulation. In this study, our attention is focused on the average concentrations of SF6 in the closed room and the spatial and temporal distribution of the concentration of SF6. Air velocity flow and temperature are also part of the reserved results (**Tables 2 and 3**).

Type of grid	Number of nodes
Dense mesh (DM)	3,218,400
Less dense mesh (LDM)	1,327,500
Coarse mesh (CM)	442,500

Table 2.
Number of nodes for each mesh.

Equipment	Positions (m)					
	Xmin	Xmax	Ymin	Ymax	Zmin	Zmax
Bed	1.40	3.20	2.05	2.75	0.0	0.40
Patient	1.50	3.15	2.25	2.55	0.40	0.65
Door	0.15	1.05	−0.02	0.00	0.05	2.00
Lamp 1	0.40	1.00	0.40	1.00	2.50	2.50
Lamp 2	2.20	2.80	0.40	1.00	2.50	2.50
Lamp 3	2.20	2.80	2.10	2.70	2.50	2.50
Lamp 4	0.40	1.00	2.10	2.70	2.50	2.50
Lamp 5	2.20	2.80	4.20	4.80	2.50	2.50
Lamp 6	0.40	1.00	4.20	4.80	2.50	2.50
Plaque 1	2.40	3.00	1.05	1.65	2.40	2.43
Plaque 2	2.40	3.00	3.65	4.25	2.40	2.43
Supply diffuser 1 (scenario 1) and (scenario 2)	2.40	3.00	1.05	1.65	2.50	2.50
Supply diffuser 2 (scenario 1) and (scenario 2)	2.40	3.00	3.65	4.25	2.50	2.50
Exhaust grille 1 (scenario 1) and supply diffuser 1 (scenario 3)	1.00	1.60	3.65	4.24	2.50	2.50
Exhaust grille 1 (scenario 1) and supply diffuser 2 (scenario 3)	1.00	1.00	1.00	1.60	2.50	2.50
Exhaust grille 1 (scenario 2) and (scenario 3)	2.40	2.90	4.80	4.80	0.40	1.00
Exhaust grille 2 (scenario 2) and (scenario 3)	2.40	2.90	2.40	2.40	0.40	1.00

Table 3.
Component positions of modeled room, supply diffuser, and exhaust grille in each scenario.

The results of scenario 1 are presented in **Tables 4** and **5**. As can be seen, the tables show a comparison between concentrations simulated by FDS and those simulated by Cheong and Phua [1]. The concentration simulated by FDS [C_{FDS}] and the experimental concentrations obtained by Cheong and Phua [1] [C_{exp}] are presented, respectively, in **Table 4**.

Error expressed as a percentage is given by $|([C_{FDS}] - [C_{exp}])/[C_{exp}]|$ for the concentration and by $|([V_{FDS}] - [V_{ex}])/[V_{ex}]|$ for the air velocity.

The concentrations obtained at points SP1, SP2, SP3, and SP5 by the LES code are inferior to the experimental results of Cheong and Phua [1]. These low concentrations can partly be explained by the infiltration of air under the door. Although Cheong and Phua [1] do not specify this infiltration rate or the pressure difference between the corridor and the closed room, AIA recommends that, for an operating room maintained at negative differential pressure, air flow to the extraction outlets must be 10% higher than the permitted air flow rate. In this case, the flow rate is

	SP1	SP2	SP3	SP4	SP5
[C_{FDS}]	20.00	22.50	20.40	34.20	28.60
Cheong and Phua [C_{exp}]	28.9 ± 0.7	28.0 ± 0.5	28.2 ± 0.6	33.4 ± 1.7	32.9 ± 0.9
Error (%)	31	20	28	2	10

Table 4.
SF6 concentration (ppm) simulated by FDS and Cheong and Phua [1].

	SP1	SP2	SP3
V_{FDS} (m/s)	0.15	0.15	0.12
V_{ex} (m/s)	0.14 ± 0.1	0.18 ± 0.2	0.16 ± 0.2
Error (%)	7	17	25

Table 5.
Air velocity simulated by FDS and Cheong and Phua [1].

14% higher than the blowing rate. Since the infiltration rate is slightly higher than recommended, the gaseous contaminant dilution will tend to be more efficient. This will lead to an average concentration of SF6 in the room, which is inferior to that obtained at a lower infiltration rate.

Table 5 shows both the simulated and experimental flow velocity modules, expressed in m/s. The V_{FDS} represents the average air velocity by the FDS over a range of 200 s (800–1000 s). In contrast, V_{ex} is the experimental air velocity in Cheong and Phua's [1] study. As can be seen, there is excellent correlation between what is simulated by both methods and what is simulated by Cheong and Phua [1]. The deviations, shown as percentages, appear to be high and are expressed in cm/s. Specifically, they are less than 5 cm/s, which is not significant.

3.2 Scenarios 2 and 3: results

The concentrations obtained for scenarios 2 and 3 are presented in the following table.

Although FDS predicts a concentration that is slightly lower than the numerical code used by Cheong and Phua [1], the results obtained by FDS in scenario 2 are in agreement with those estimated by the researchers. On the other hand, considerable error is observed for scenario 3, in which the room is supplied with fresh air by two grids located on the ceiling. These grids tend to force the air to the floor. Points SP1 and SP3 are directly under the grids, and the concentration of SF6 is very low (<1 ppm). In this context, it is surprising that Cheong and Phua [1] obtained a concentration of 29.0 ppm at these points, as the only possible explanation would be related to the position of the supply grids. Although a plan of the room in Cheong and Phua's [1] article seems to indicate that the supply grids are located directly under points SP1 and SP3, there is no information on the exact positioning of the grids (Tables 6 and 7).

Moreover, Cheong and Phua [1] do not give the recommended air temperature for the room. If the supply air temperature is high, then the Archimedes thrust will tend to significantly decrease the range of the jet. The influence of the blowing

Position	Scenario 2			Scenario 3		
	C_{FDS}	Cheong and Phua. [C.num]	Error (%)	C_{FDS}	Cheong and Phua. [C.num]	Error (%)
SP1	32.0	29.0	10	~0	29.0	10.0
SP2	28.4	34.0	18	21.2	28.0	24.0
SP3	27.3	30.5	11	~0	29.0	100.0

Table 6.
Numerical and simulated concentration (ppm) results.

Position	Scenario 2		Scenario 3	
	PREFDS	PRECheong and Phua	PREFDS	PRECheong and Phua
SP1	1.00	1.08	∞	1.08
SP2	1.13	0.91	1.22	1.12
SP3	1.17	1.03	∞	1.08

Table 7.
Pollutant removal efficiency.

temperature was checked through the increase by 2°C in the air temperature admitted into the room. The concentrations obtained at SP1 and SP3 remained negligible, which indicates that much higher temperatures would be required to reduce the range of the jets to a few tens of cm.

The flow velocity achieved a value of 0.05 m/s in nearly the entire room, except for the floor, where the velocity reached a value of 0.8. At the level of the diffusers, the air velocity was 0.6 m/ s. Further, it was observed that the jet of air coming from the corridor faded before reaching the other end of the room. This jet of air diffused in the vertical direction, which helped to dilute the gaseous contaminant in the room.

The influence of the blowing grids on the concentrations is clearly visible. As mentioned earlier, very significant changes in concentration are observed in the area above the patient's bed. A slight change in the position of the blown grids is likely to have a significant impact on the simulated concentrations at SP1 and SP3. Point SP2, which is situated at the foot of the patient's bed, is in an area where variations in concentration are less important.

3.3 Pollutant removal efficiency index

In order to compare the effectiveness of the three ventilation strategies, the pollutant removal efficiency (PRE) index is used.

PRE is calculated as follows:

$$PRE = \frac{C_s}{C_j} \qquad (7)$$

where C_s is the average concentration of SF6 in the exhaust air grille; C_j is the average concentration of SF6 at points (x, y, z).

The PRE can be calculated for a room with more exhaust grilles by averaging the concentrations obtained at the various extraction grids. The PRE index is used to quantify the effectiveness of ventilation to remove pollutants from a room. It depends on a number of factors, such as the location of the source of the pollutant, the supply flow, the ventilation strategy, etc. The strategy is effective for removing pollutants and represents a good ventilation solution if PRE > 1. On the other hand, if PRE < 1, there is an accumulation of contaminants in the room. This could be related to, for example, the existence of recirculation zones where the contaminants accumulate.

For scenario 2 in the three points of SP1, SP2, and SP3, we obtain PRE > 1. This result is the same as that in Cheong and Phua's [1] work, except for SP2. In scenario 3, the FDS simulation gives completely different results than Cheong and Phua's [1] for the two points of SP1 and SP3.

4. Conclusions

This paper investigated the air quality in a bronchoscopy unit. The numerical model used in the present study was based on the large-eddy simulation (LES) method. The numerical results obtained in this work have been generally validated by the experimental results found in the literature. For this investigation, we used Cheong and Phua's [1] study for validation and comparison purposes. Three numerical scenarios (scenario 1, scenario 2, and scenario 3) were developed according to ASHRAE norms and standards. Fire Dynamics Simulator software was used to estimate the concentration of contaminants and the air velocity in the bronchoscopy unit. According to the results obtained, both scenario 1 and scenario 2 are effective for removing SF6 (pollutants). However, according to our results, scenario 3 should not be retained, as in this scenario, the concentration of pollutants was very high compared to the other two scenarios. Moreover, the concentration of SF6 accumulated around the patient's bed.

In light of these findings, the authors of the present work suggest that more in-depth investigation into the air quality of hospitals is warranted. This could be combined with field experiments using scenarios 1 and 2.

Author details

Hanaâ Hachimi[1], Chakib El Mokhi[2], Badr T. Alsulami[3] and Abderrahim Lakhouit[4*]

1 Sultan Moulay Slimane University, Beni Mellal, Morocco

2 National School of Applied Sciences, Ibn Tofail University, Kenitra, Morocco

3 Department of Civil Engineering, College of Engineering and Islamic Architecture, Umm Al-Qura University, Saudi Arabia

4 Civil Engineering Department, Faculty of Engineering, University of Tabuk, Saudi Arabia

*Address all correspondence to: a.lakhouit@usherbrooke.ca

References

[1] Cheong KWD, Phua SY. Development of ventilation design strategy for effective removal of pollutant in the isolation room of a hospital. Building and Environment. 2006;41(9):1161-1170

[2] Sundell J. On the history of indoor air quality and health. Indoor Air. 2004;14 (s7):51-58

[3] Kuehn TH. Airborne infection control in health care facilities. Transactions-American Society of Mechanical Engineers Journal of Solar Energy Engineering. 2003;125(3): 366-371

[4] Dales R, Liu L, Wheeler AJ, Gilbert NL. Quality of indoor residential air and health. Canadian Medical Association Journal. 2008;179(2): 147-152

[5] Jones AP. Indoor air quality and health. Atmospheric Environment. 1999;33(28):4535-4564

[6] Sautour M, Sixt N, Dalle F, L'Ollivier C, Fourquenet V, Calinon C, et al. Profiles and seasonal distribution of airborne fungi in indoor and outdoor environments at a French hospital. Science of the Total Environment. 2009; 407(12):3766-3771

[7] Lakhouit A. Modélisation de la qualité de l'air dans une unité de bronchoscopie: influence des stratégies de ventilation. 2011. Thèse de doctorat. École de technologie supérieure

[8] Saad SG. Integrated environmental management for hospitals. Indoor and Built Environment. 2003;12(1–2):93-98

[9] De Giuli V, Zecchin R, Salmaso L, Corain L, De Carli M. Measured and perceived indoor environmental quality: Padua hospital case study. Building and Environment. 2013;59:211-226

[10] Pittet D, Allegranzi B, Sax H, Bertinato L, Concia E, Cookson B, et al. Considerations for a WHO European strategy on health-care-associated infection, surveillance, and control. The Lancet Infectious Diseases. 2005;5(4): 242-250

[11] Hathway A, Noakes CJ, Sleigh PA. CFD modeling of a hospital ward: Assessing risk from bacteria produced from respiratory and activity sources. In: Indoor Air 2008: The 11th International Conference on Indoor Air Quality and Climate. Leeds; Indoor Air 2008, 17-22nd August 2008, Copenhagen, Denmark

[12] Qian H, Li Y. Removal of exhaled particles by ventilation and deposition in a multibed airborne infection isolation room. Indoor Air. 2010;20(4): 284-297

[13] McGrattan K, Hostikka S, McDermott R, Floyd J, Weinschenk C, Overholt K. Fire Dynamics Simulator, User's Guide. 6th ed. National Institute of Standards and Technology: NIST Special Publication; 2013. p. 1019

[14] Barrero D, Hardy J-P, Reggio M, Ozell B. CFD and realistic visualization for the analysis of fire scenarios. In: ACM SIGGRAPH 2004 Posters. ACM; 2004. p. 101

[15] Temam R. Navier-Stokes Equations. Vol. 2. North-Holland Amsterdam; 1984

[16] Persily A. Challenges in developing ventilation and indoor air quality standards: The story of ASHRAE standard 62. Building and Environment. 2015;91:61-69

Chapter 3

Prediction of Agricultural Contaminant Concentrations in Ambient Air

Steven Cryer and Ian van Wesenbeeck

Abstract

Monitoring ambient air to assess environmental exposure and risk for volatile agricultural chemicals requires extensive resources and logistical effort. The cost and technical limitations of monitoring can be mitigated using a validated air dispersion model to simulate concentrations of volatile organic chemicals in ambient air. The SOil Fumigant Exposure Assessment (SOFEA) model was developed to explore volatile pesticide exposure and bystander risk. SOFEA assembles sources and source strengths, uses weather data from the region of interest, and executes an air dispersion model (AERMOD, ISCST3) to simulate pesticide concentrations at user defined receptors that can be used in exposure and risk assessment. This work highlights SOFEA development from inception and modifications over the last 1.5 decades, to the current delivery within the public domain. Various examples for the soil fumigant 1,3-dichloropropene are provided.

Keywords: air dispersion modeling, SOil Fumigant Exposure Assessment tool (SOFEA), AMS/EPA Regulatory Model (AERMOD), 1,3-dichropropene, Gaussian plume

1. Introduction

The development of a numerical modeling tool for the soil fumigant 1,3-D started several decades ago using the Industrial Source Complex Short Term (ISCST3) air dispersion model [1]. Early work was extended by incorporating a soil fate modeling tool, the Pesticide Root Zone Model (PRZM3), to simulate the source strength used in ISCST3 air dispersion calculations [2]. This initial work was the forerunner of the SOil Fumigant Exposure Assessment system (SOFEA), a stochastic numerical modeling tool developed by Corteva Agriscience as a regulatory tool to evaluate and manage human inhalation exposure potential associated with the use of soil fumigants and other semi-volatile or volatile compounds [3]. There are no existing models for predicting pesticide exposure that can easily incorporate multiple fields throughout the year that mimic use rates and volatility that ulti- mately govern exposure. Even today, SOFEA has more attributes and functionality when addressing exposure risk from the use of volatile (or semi-volatile) pesticides than other agricultural models.

SOFEA calculates fumigant concentrations in air arising from volatility losses from treated agricultural fields for entire agricultural regions using multiple transient source terms (treated fields), Geographic Information System (GIS) information, agronomic specific variables, user specified buffer zones, and field re-entry intervals. The United States Environmental Protection Agency (USEPA) Gaussian plume models ISCST3 [4] and/or the AMS/EPA Regulatory Model (AERMOD) [5–6] are used for air dispersion calculations. Examples of ISCST3 simulations include modeling vehicle exhausts in urban areas [7] and modeling the ambient air concentrations at specified buffer distances for single fields in agriculture [8]. AERMOD simulations include estimates of mercury levels in air [9]), NO_2 and SO_2 predictions in Thailand [10], and air concentrations resulting from emissions from agricultural fields treated with the soil fumigant 1,3-dichloropropene [11].

This work provides a summary for SOFEA development and use in agriculture over the past decade. SOFEA uses field observed (or numerically generated) fumigant flux profiles from soil as transient source terms for both agricultural shank injection and drip-irrigation applications of a soil fumigant. Measured reference flux observations are scaled based upon depth of incorporation into soil and the time of year, to map the complete flux response surface from field/numerical observations, however a soil physics model can also be used to estimate flux from soil (e.g., HYDRUS [12], CHAIN_2D [13], etc.). Weather information, field size, application date, application rate, application type, depth of soil incorporation, pesticide degradation rates in air, tarp presence at the soil surface, ag-capable land, field re-treatment from 1 year to the next, buffer setbacks, and other sensitive parameters are varied stochastically using Monte Carlo techniques to mimic region and crop specific agronomic practices. Agricultural regions up to 49,210 km^2 (19,000 mi^2) can be simulated for temporal periods ranging from 1 day to more than 70 years for the purpose of addressing acute (24 h), short-term (72 h), sub-chronic (28 or 90 days) or chronic exposure. Multi-year, multiple field simulations are conducted using either random field placement in all agricultural capable areas, by selectively placing fields in historical or prospective areas, and/or placing fields of a specific size and spatial location if this information is available. Regional land cover, elevation, and population information cavn be used to refine source placement (treated fields), dispersion calculations, and exposure assessments. Both current and anticipated/forecasted fumigant scenarios can be simulated to provide risk managers with the necessary information to make sound regulatory decisions.

SOFEA has been used for regulatory decision making in California, was reviewed in the 2004 USEPA Scientific Advisory Panel (SAP) meeting [14], and is currently being used in the registration review process for 1,3-dicholopropene (1,3-D) with USEPA. Algorithms used by SOFEA to refine exposure predictions and manage acute, sub-chronic, and chronic risk associated with the use of soil fumigants on a local or regional basis are presented. Although SOFEA was originally designed specifically to describe air concentrations for the soil fumigant 1,3-D, the model is readily adaptable to generically describe the post-fumigation air concentrations of other soil fumigants and semi-volatile organic chemicals. SOFEA can now be executed in "retrospective" mode which allows the user to specify specific field locations where known fumigant applications are made and specific receptor locations where fumigant concentrations are measured which makes the model predictions available for comparison to field monitoring observations. Other enhancements to SOFEA include the option of importing flux (emissions) from soil simulated by a soil physics model (HYDRUS, CHAIN_2D) in lieu of flux obtained from field experiments. SOFEA will soon be in the public domain and can be obtained from Exponent Inc. (U.S. based consulting company) for use with all volatile or semi-volatile pesticides if parameterized appropriately. This manuscript summarizes SOFEA capabilities in both in prospective and retrospective mode, from inception to release of the model to the public domain.

2. Background

A generic methodology to determine fumigant concentrations in ambient air in large and diverse air sheds has been developed (**Figure 1**). Directionally averaged air concentrations within entire air sheds are determined using a multiple source Gaussian dispersion model that has been modified to include Monte Carlo sampling techniques, ties to GIS databases, and agronomic practices. Time averaged transient air concentrations simulated via a numerical model can be used to assess exposure and risk for an unlimited number of scenarios.

SOFEA enables the determination of "area-wide" concentration profiles for user specified distances that account for multiple field applications. Thus, the effect of fields "off-gassing" at different points in time and space are accounted for by SOFEA. The user can evaluate the impact of the buffer on the acute exposure for residents and by-standers by specifying a buffer distance from the edge of treated fields. The user can also determine the chronic exposure to individuals residing in the use area by specifying the total mass applied on an annual basis and running SOFEA for a full year or multiple years. SOFEA inputs and outputs are easily exported to other file formats or programs. Concentrations of soil fumigants in air

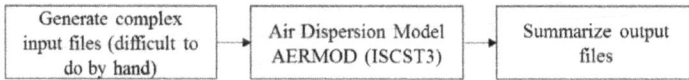

Figure 1.
SOFEA is an intelligent input file generator and output repository for agronomic use of the USEPA Gaussian plume model AERMOD (and formerly ISCST3).

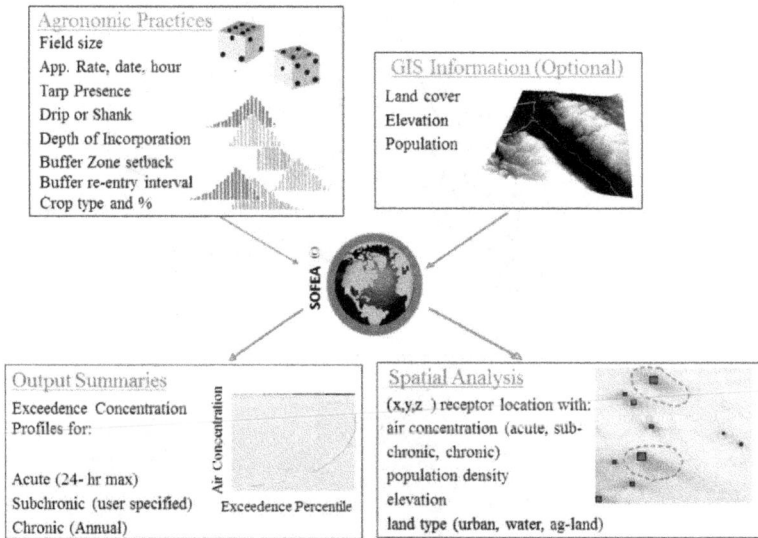

Figure 2.
SOil Fumigant Exposure Assessment Model overview.

are associated with x, y, z co-ordinates and associated with proximity to treated fields as well as human populations (if population census data is available). The SOFEA model is readily adaptable to generically describe the post-fumigation air concentration of other organic contaminants, **Figure 2**.

3. SOFEA refinements and use, past and present

3.1 Air dispersion model

The ISCST3 [4] and AERMOD [5, 6] models were developed by the USEPA as regulatory tools for predicting concentrations of air contaminants in diverse air sheds. Both are Gaussian plume models useful for estimating air quality surrounding contaminant release sites. AERMOD replaced ISCST3 by USEPA, although ISCST3 is still in use by some researchers and regulatory authorities. ISCST3 has been widely used to simulate probability distribution functions (PDFs) of fumigant concentrations in air within townships in California to estimate acute and chronic bystander exposure [15, 16].

3.2 Parameter representation

The complex terrain algorithms of ISCST3/AERMOD can account for the effects of elevation changes within specific regions should this information be available. Population information (if provided or known) can also be used in population-based risk assessments. The 2010 U.S. census data lists population densities by census blocks and is a good choice for population information. A township is defined

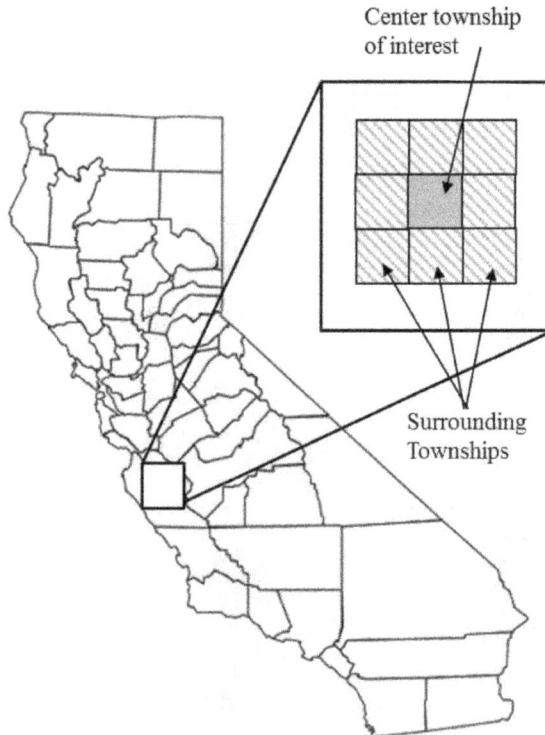

Figure 3.
Example of a 3 × 3 township air shed for SOFEA simulation from the state of California.

according to the Public Land Survey System (PLSS) and is nominally 6 × 6 mi (9.66 × 9.66 km) in area, **Figure 3**. The spatial locations for receptors placed uniformly or weighted in a central township are user specified, and appropriate land cover, elevation, and population data from GIS data bases are necessary inputs. Township information must include land cover such that ag-capable land can be quantified. Elevation and population information are optional. The impact of sources external to the central township domain will depend on the persistence and drift characteristics of the pesticide being simulated. However, a model evaluation for 1,3-D showed that 3 × 3 townships was adequate to account for edge effects [17].

3.3 Air shed simulation domain

An air shed is defined as a volume of air overlying a square surface area, where source terms (i.e., treated fields) throughout the air shed can contribute to overall air concentrations at specific locations. Although historical SOFEA simulations focused on air concentrations in either a single township or a 3 × 3 township domain, the model can be used to simulate concentrations across much larger airsheds. The complex terrain algorithms of ISCST3 or AERMOD can take advantage of elevation changes within specific regions. Receptors can be placed uniformly in a central township or over a multi-township domain. Source terms can be placed anywhere in the simulation domain, which can include up to 23 × 23 townships (49,210 km^2 = 19,000 mi^2). When running in prospective mode, the user need only specify the annual pesticide mass applied to any township within a 23 × 23 township domain, appropriate GIS information, receptor spacing and heights, as well as appropriate PDFs characterizing agronomic practices within the region.

3.4 Stochastic portrayal

Concentrations of a soil fumigant in air resulting from transient agricultural source terms are also dependent upon meteorological conditions, application timing, and other agronomic properties. A mechanism was required that could propagate parametric uncertainty in sensitive model inputs to air concentration predictions. Monte Carlo (MC) methods provide a straightforward technique to propagate such uncertainty in independent parameters to dependent output variables [18, 19]. PDFs can include fumigated field sizes, application rates, application dates etc. SOFEA can also be used in retrospective mode, where exact treated field locations and application parameters (mass applied, date applied, etc.), and receptor locations are known. Variability in input is described by PDFs that are randomly sampled to generate input parameter sequences. Stochastic variables for SOFEA include application rate, date, and hour of day initiated, pesticide depth of incorporation, presence of a tarp at the soil surface, application type (shank injection or drip irrigation), field size, weather year, and pesticide properties such as degradation rates in air. Output predictions are no longer deterministic, but rather a discrete distribution is generated from which exceedance probabilities and return frequencies can be calculated (e.g., 1-in-100-year exposure potential, and so on). Air quality modeling work is in accordance with the policy established by the U.S. EPA for Air Quality Models and follows the guidelines set forth by U.S. EPA for Monte Carlo Analysis [20].

The original version of SOFEA required the MS Excel add-on program Crystal Ball™ (Decisioneering, Inc.) to transform ISCST3 from a deterministic model into a stochastic/deterministic system; however, subsequent versions have been modified to include Visual Basic Applications (VBA) algorithms that obviate the need for Crystal Ball™. In SOFEA2, an ISCST3 input file is exported from Excel that is based upon appropriate selections from user defined PDFs that are derived from actual agronomic data. Excel, ISCST3, and VBA programs were coupled to allow the transparent integration of the Monte Carlo component in SOFEA3, which used, but ISCST$_3$ had changes in mixing height calculations under calm conditions such that simulation with parametric uncertainty more closed matched monitoring observations [21]. The latest version of SOFEA (SOFEA4) contains identical functionality as the original version but was rewritten in C++ and Qt to replace VBA programming, ISCST3 replaced by AERMOD, and results using SOFEA4 use are found elsewhere [11]. SOFEA4 provides automation, transparency, a Graphical User Interface (GUI), the use of AERMOD, and maintainability for future support.

3.5 Crop selection and simulation domain

Fumigants are used on a variety of high valued agricultural commodities. Each commodity/crop is potentially unique, with different agronomic management practices. The crops chosen can be based upon current or future forecasted fumigant uses, and currently up to five different crop types can be considered. Predominant crops where soil fumigants are used include tree and vine (TV), field crops (FC), nursery crops (NC), strawberries (SB) and post-plant vines (PP). The contributions of a soil fumigant to air quality from each crop are easily extractable by keeping the crop types/parameters unique during simulation. This aids in determining appropriate Best Management Practices (BMP's) by crop type. SOFEA uses the supplied PDFs to generate the agronomic variables (e.g., field size, application rate, etc.) for each crop type. Thus, if a region is dominated by one crop type, all five crop types in SOFEA can be parameterized with the same data (if desired) to minimize computer memory requirements.

3.6 Receptors

Receptors are specific (x, y, z) locations in the simulation domain where air concentrations are calculated. These receptors can be uniformly placed within the township for chronic exposure predictions, or at specific setback distances around treated fields if acute exposure assessment is required, **Figure 4**. Historical SOFEA simulations in CA have assumed a rectangular grid of 36 equally spaced receptors per township section (i.e., a 1 mi^2 area) which yields 1296 receptors per township at a spacing of 268.2 m (880 feet) [11, 21]. Receptors are typically placed at 1.5 m above the ground to mimic the breathing height on adult. Ultimately, any desired

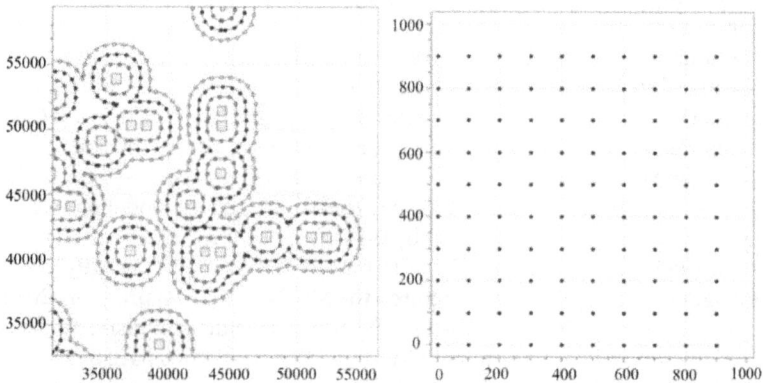

Figure 4.
Receptor placement at user specified buffer setback near treated fields (left) or uniform grid placement (right).

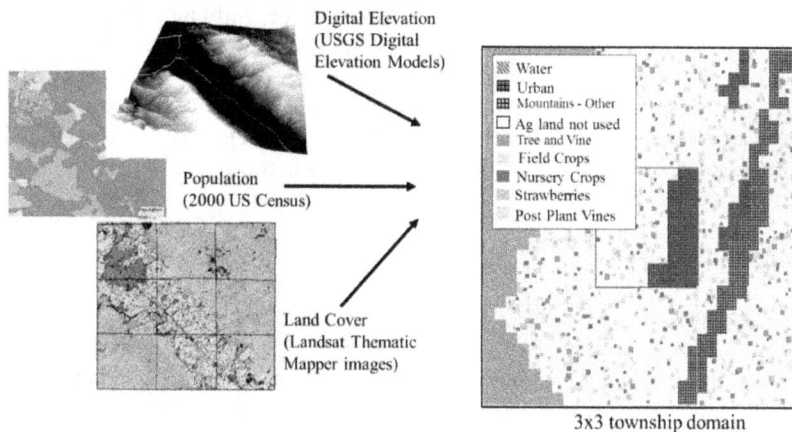

Figure 5.
Example of the continually updated public domain data bases available to the user of SOFEA.

receptor location and density, and height can be specified by the user, allowing individual receptors to be placed anywhere in the simulation domain.

3.7 GIS data layers

Many data bases and GIS software programs exist to extract appropriate information for use in SOFEA, **Figure 5**. SOFEA is not a GIS tool but rather uses GIS information that has been assembled using software such as ArcView™ (ESRI, Inc.). Land cover information is obtained by Landsat Thematic Mapper images (30-m resolution) that contain 21 unique land classifications [available from the National Land Cover Data (NLCD) database at http://landcover.usgs. gov/natllandcover.htm]. Elevation information is obtained from the USGS Digital Elevation Models (DEM) data at 1:24,000 scales. Population information is given by census blocks and populated with data from the 2010 US Census, and GIS information is used to parameterize the air shed for ISCST3/AERMOD simulations (e.g., ag-capable land where fields can be placed, etc.).

3.8 Meteorological data

A single location for weather data is used in SOFEA to represent weather conditions from the region of interest. Meteorological information includes hourly air stability class, wind speed, air temperature, wind direction and mixing and ceiling height for ISCST3, along with the Monin-Obukhov stability length for AERMOD. Wind speed and direction are critical parameters, and for larger simulation domains with potentially greater surface roughness length (z) (due to trees, buildings, fence rows, etc.), wind speed is preferable measured at a height of 10-m. A rule of thumb for determining the minimum height of the wind sensor is $7*z$ [22]. Flat fallow fields typically have a roughness length $z < 0.1$-m, and therefore an anemometer height of 2-m is adequate. The user creates a weather library for each year of weather and this library is assigned a uniform distribution when SOFEA is executed in prospective mode, or actual weather information for a specific time frame when running in retrospective mode. Weather data is available from public sources such as the California Irrigation Management System (CIMIS), the National Oceanic and Atmospheric Administration (NOAA), or the Florida Automated Weather Network (FAWN), or could be collected by a dedicated weather station installed in the simulation domain. The weather station should collect, at minimum, hourly precipitation, solar radiation, air tempera- ture, and wind speed and direction (SOFEA requirements). Weather data must be pre-processed using PCRAMMET if the ISCST model is used, or the AERMET pre-processor if AERMOD is used. Pre-processing is conducted outside of the SOFEA model framework.

3.9 Source placement

In prospective mode, sources (treated fields) can be placed randomly or weighted to specific township locations, **Figure 6**. All ag-capable land (all land excluding urban areas, water bodies, barren, rock, quarries, and wetlands) is used and placement is based on a uniform probability of occurrence (known as random field placement). However, there are situations in high pesticide product use regions where treated field locations are known, and section weighting can be used to ensure that product use spatially represents historical needs. A town- ship section is 1/36 of the township area and the user can specify the probability that these sections are locations where fields are placed. Receptors in these regions will register higher chronic soil fumigant air concentrations due to the increased field (i.e., source) density. Section-weighting probabilities can be based on expert judgment and/or historical product use records. When sections "fill up" and can no longer contain another treated field, a "spill-over" algorithm is introduced in SOFEA where the fields are then placed in sections surrounding the section that is "filled."

3.10 Township allocation of fumigant mass

In California, the amount of 1,3-D applied annually cannot exceed a mandated township allocation which is set based on acceptable levels of chronic exposure. Each township is assigned an allocation amount based on CA permit conditions (or some other a user-supplied amount), so this system can and has been applied in

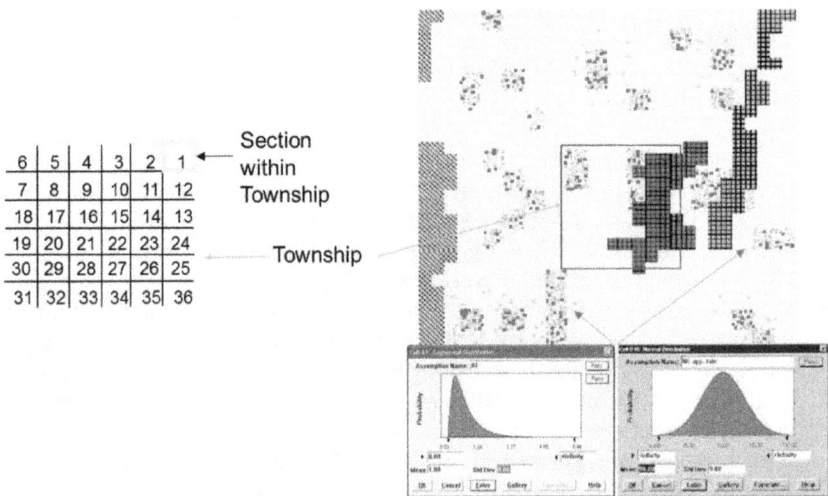

Figure 6.
Use of section weighted placement of treated fields within a township based upon specific sections having high fumigant use. Each field is assigned input based upon user specified PDFs.

other fumigant use areas across the United States. The amount of pesticide used in a given township is thus given as a fraction of this user specified township allocation.

3.11 Application scaling factor (CA only)

The California Department of Pesticide Regulations (CDPR) uses a simple procedure to account for seasonal and incorporation depth variability on pesticide volatility losses to represent the complete flux response profile. Volatility losses are sensitive to temperature and depth of soil incorporation [2] and a simple expression is used where the chemical flux from soil to the air is defined as.

$$\text{Flux}_i = R \times \text{Fr}_i \times S_{incorp} \times S_{yr} \tag{1}$$

where, Flux_i = scaled hourly flux loss from soil into air for hour "i" based upon an actual field trial, Fr_i (kg ha^{-1} h^{-1}), R = pesticide application rate (kg ha^{-1}), Fr_i = observed flux rate (reference profile based on a field experiment, or modeling), S_{incorp} = scaling factor for depth of incorporation (dimensionless), and S_{yr} = scaling factor for time of year (dimensionless).

Although the CDPR approach only uses a single flux profile for each application type, these profiles are modified by soil incorporation depth and time of year. Also, models such as HYDRUS [23], STANMOD [24], CHAIN_2D [13] and PRZM3 [25] can also be used to develop flux profiles for different conditions.

3.12 Temporal representation S_{yr} (CA only)

California is sectioned into warm and cool seasons where increased emission to the atmosphere occurs under warm conditions and is arbitrarily increased by a factor of 1.6× by CDPR. Therefore, S_{yr} is assigned a value of 1.6 to account for gross seasonal temperature effects during the warm season. This warm season can be a specific time of the year (as for CDPR) or the SOFEA user can use a continuous sinusoidal function, where the amplitude and frequencies are daily average air temperatures based upon what day within the year a pesticide application is made (**Figure 7**, left). In addition, several constraints on the depth of incorporation are used for CA and are given in **Figure 7** (right). The user can specify how the pesticide incorporation depth in soil can alter the cumulative mass loss from the soil surface (linear, exponential, CDPR) by selecting from options in the drop-down

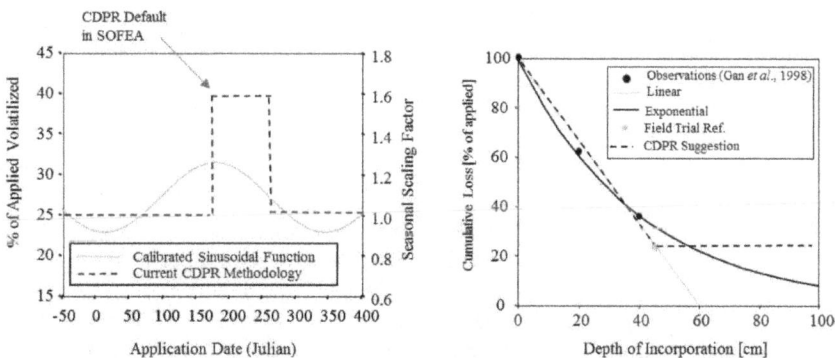

Figure 7.
Example of CDPR application factor or sinusoidal modeling for percent of applied volatilized (left) and impact of depth of pesticide incorporation (right).

Meteorological Data and Agronomic
Practices

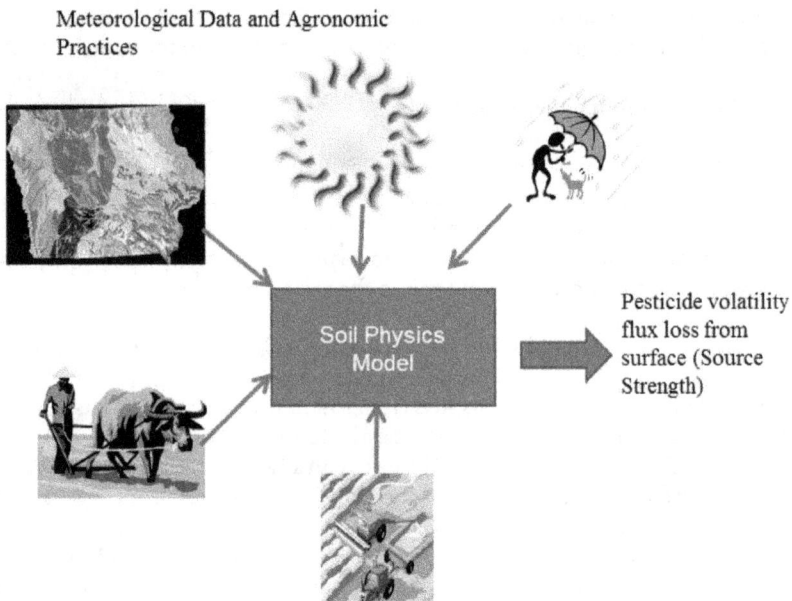

Figure 8.
Obtaining soil volatility flux estimates using a model such as CHAIN_2D or HYDRUS.

menu in the SOFEA GUI. The type of seasonal scaling (CDPR or sinusoidal) can also be selected by the user in the SOFEA GUI.

3.13 Model output characterization

SOFEA is used to execute the air dispersion models ISCST3 (historical) and/or AERMOD (recent). Hourly output from these models can be analyzed according to user selections for post processing output concentrations (e.g., 1-h, 1-day, 3-day, 15-day, annual, and so forth). Functionality for the current version of SOFEA is somewhat different than in the earlier versions, but the bulk of functionality for SOFEA are the same and found elsewhere [3].

In earlier versions of SOFEA, input and output were facilitated via a VBA interface that utilizes EXCEL spreadsheets containing user supplied PDF's of application parameters. Users could create inputs based on actual field data and pesticide use information, or generate hypothetical distributions of use parameters such as field sizes, application rates and timing, depth of injection, etc. Over the years, SOFEA has evolved from a VBA model using only ISCST3 to a C++ interface that can drive AERMOD simulations (e.g., SOFEA4). A user guide for the most recent version of SOFEA4 is currently in preparation and should be available sometime in 2019.

Chemical flux estimates can be obtained from a variety of different experimental sources but can also be estimated from soil physics models such as HYDRUS [23], STANMOD [24] and CHAIN_2D [13], **Figure 8**. Such models can and have been used to simulate both volatility from soil along with movement into the soil profile. Advantages of using a soil physics model to estimate pesticide flux loss deal with low cost and the semi-infinite parameter space that can be explored with simulation techniques. Field studies in five different states that explored atmospheric flux loss for chloropicrin and 1,3-dichloropropene were validated with CHAIN_2D and indicate the model can correctly capture both peak and cumulative emissions effectively for these two soil fumigants [17, 26]. Thus, CHAIN_2D and similar models are useful

Soil Column
(1/2 plane)

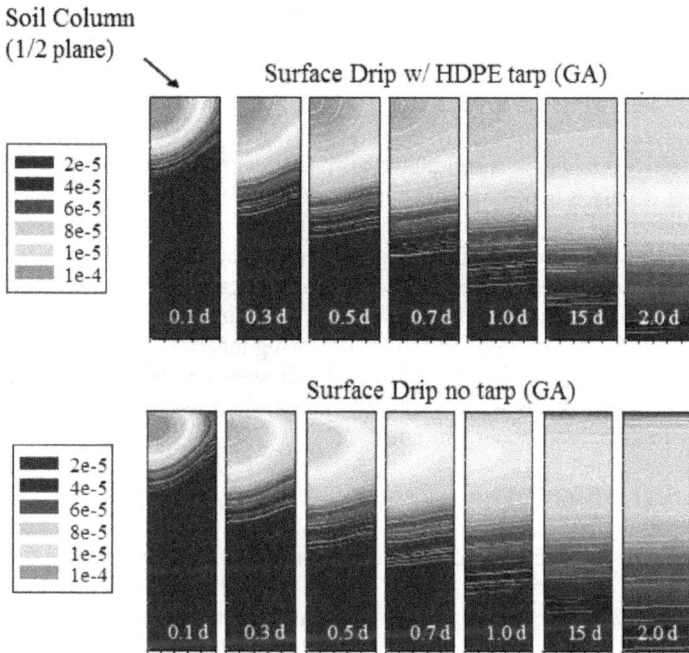

Figure 9.
Example use of soil physics model CHAIN_2D where exploration of multiple mitigations strategies such as water sealing, tarps, depth of incorporation, etc. can be simulated, as well as concentration by time (C × T) within soil for biological efficacy.

tools for extrapolating flux predictions to diverse scenarios where experimental observations are unavailable [27–29]. Examples of various mitigations strategies that can be explored using soil physics models include the use of agricultural films, increased soil injection depth for the fumigant, and under a near semi-infinite parameter combinations of meteorological, soil and agronomic properties, **Figure 9**.

Soil physics models should first be validated with field observations before being used to extrapolate to a variety of different conditions. Cryer and van Wesenbeeck [26] used CHAIN_2D to validate against field observations, and then coupled CHAIN_2D to several USEPA air dispersion models (ISCST3 [4] and CALPUFF [30]). Both cumulative and 1-h maximum air concentrations were simulated and compared against field observations with good success (the best observations and simulations results were between 6 and 8%). In addition, both ISCST3 and CALPUFF air dispersion models showed similar order of magnitude output predictions [17]. Chloropicrin and 1,3-D emissions through Totally Impermeable Film (TIF) were compared using HYRUS where the fumigant flux was simulated within a factor of ~2, though the timing of the peak was over-predicted by the model [29]. The authors suggest that field-based calibration should be conducted when tarps are used because of the lack of representative field effective permeability data for the tarps.

Most inputs can be specified as either discrete values, or as PDFs. If possible, PDFs should be used to maximize Monte Carlo capabilities of the SOFEA modeling system and encompass uncertainties and variability in model inputs. SOFEA can generate fumigant concentrations for each receptor in the simulation domain (up to 11,664 receptors have been simulated in a nine-township air shed), averaged over specific time intervals (24-h and yearly) or periods specified by the user. For example, the user could specify the output of 24-h average, 60-day average, and annual average concentration PDFs, for assessing acute, sub chronic and chronic risk to exposed populations.

3.14 SOFEA sensitivity analysis

A sensitivity analysis was conducted with earlier versions of SOFEA to determine which variables had the greatest impact on model predicted concentrations [3]. The dependent variable endpoint in the sensitivity analysis for Kern County, CA was the 15-D multi-direction average air concentration at 30.5 m buffer. Sensitive parameters were the crop percentage, application rate, application date and weather year, in addition to the amount of 1,3-D mass applied in a township and the proximity of a treated field to a monitoring location. Additional parametric sensitivity analysis for CHAIN_2D/ISCST3 showed several soil and irrigation parameters as consistently sensitive, including depth of incorporation into soil, tarp material, and initial soil water content [17].

3.15 Historical uses of SOFEA

A moderate overprediction in air concentrations was made by SOFEA when predicting regional air concentrations for Ventura and Merced counties in California [31] which included 25 contiguous townships and treated at 1.5 times the current township allocation using 1,3-D (or at maximum levels of 1,3-D used between 1999 and 2006). However, this work provided an example of how SOFEA could be used using actual agronomic practices to manage the use of soil fumigant products and long-term exposure and risk to residents located in high-use regions. This publication also discussed how high-use rural areas leading to the highest predicted air concentrations could be used in a formalized risk assessment. The observation that the high concentrations were surrounding the downwind locations around treated fields was first predicted by Cryer and van Wesenbeeck [1] before the SOFEA modeling tool was fully developed.

SOFEA was improved with the release of SOFEA2 which eliminated the need for the third-party software Crystal Ball™ while also incorporating the ability to specify unique agronomic fields and air monitoring receptor locations. Further refinement includes post-processing hourly concentration predictions for precise starting intervals, the capacity to incorporate specific field flux loss from soil physics modeling for each field in the simulation domain, and the ability to accommodate drip applications made to vineyards. SOFEA and SOFEA2 generated the same output distributions when identically parameterized (unpublished work of Corteva Agriscience).

A 1,3-D air monitoring study was conducted in a high fumigant use area in Merced, CA, where 3-day average air concentrations were measured continuously at the approximate center at each of nine townships over a 14½ month period [21]. This monitoring study was designed specifically for validating SOFEA. Although SOFEA2 predicted the general pattern and correct order of magnitude for 1,3-D air concentrations as a function of time, it failed to recover the highest observed 1,3-D concentrations of the monitoring study which typically occurred in December. It was found the atmospheric mixing height was a significant parameter affecting the modeled 1,3-D concentrations. An algorithm that adjusted the PCRAMMET mixing height based on measured wind speed and air temperature was found to improve the simulated concentrations significantly, however the inclusion of AERMOD in SOFEA3 improved the model fit to observed data without requiring any mixing height adjustment. Comparison of the output probability density functions (PDFs) for 72 h 1,3-D concentrations between monitoring observations and SOFEA4

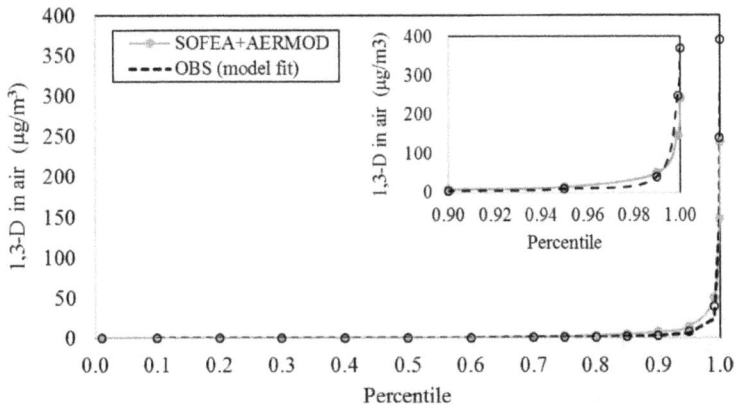

Figure 10.
Exceedance probability for Merced CA study from measured and simulated conditions when AERMOD replaced ISCST3 in SOFEA3 as the dispersion model for predictions.

simulation indicate that slight under-prediction of concentrations above the 99th percentile was off-set by slight over-prediction of the 1,3-D concentration distribution below the 99th percentile, resulting in the annual average 1,3-D concentration for the nine-receptor monitoring domain being slightly over predicted (<2%). This suggests that without further refinement, based upon field validation observations, SOFEA2 results are representative but conservative estimates of exposure for 1,3-D if border township contributions and mixing height (MH) adjustments for calm periods are considered. SOFEA2 proved a useful tool for estimating airborne levels of 1,3-D but showed some weakness when incorporating ISCST3 [21] and was renamed SOFEA3 when AERMOD was included.

AERMOD was used in conjunction with SOFEA2 after 2016 (now named SOFEA3), following the knowledge that MH was one of the most sensitive variables and that ISCST3 and its associated meteorological pre-processer over-estimated MH during stable air (calm conditions). Analysis showed SOFEA3, when using AERMOD in lieu of ISCST3 as the air dispersion model, improves the predictions of observed 1,3-D concentrations, and obviates the need for adjustments of the MHs in the pro-cessed weather file, as was required with SOFEA2. Improvements are a result of the refined algorithms in AERMOD for prediction of MH during calm conditions, based on updated understanding of Planetary Boundary Layer (BPL) dynamics, the use of the Monin-Obukhov length scale (L), and the calculation of a convective and mechan-ical MH, the latter which is used only for stable conditions (when $L > 0$). **Figure 10** shows SOFEA3 results compared to the 2010–2011 California (Merced) monitoring study, while **Figure 11** represents the same simulation predictions at a much finer resolution (11,664 receptors per township) such that contour plots for air concentra-tion can be obtained.

SOFEA3 was rewritten using C++ and Qt to replace VBA, and now this latest version is denoted SOFEA4. SOFEA4 was used to simulate 1,3-dichloropro-pene (1,3-D) concentrations in ambient air in three agricultural areas in the USA where soil fumigation is a critical aspect of pest management and crop production. The regions explored by van Wesenbeeck et al. [11] are the Pacific Northwest, the mid-Atlantic coast, and the Southeast coastal plain, **Figure 12**. The Merced, CA monitoring study served to represent the southwest region of the U.S. SOFEA4 is the latest modeling tool of SOFEA that has been modified to use AERMOD, the EPA's recommended regulatory air dispersion model, to predict short-, medium- and long-term pesticide concentrations in air resulting

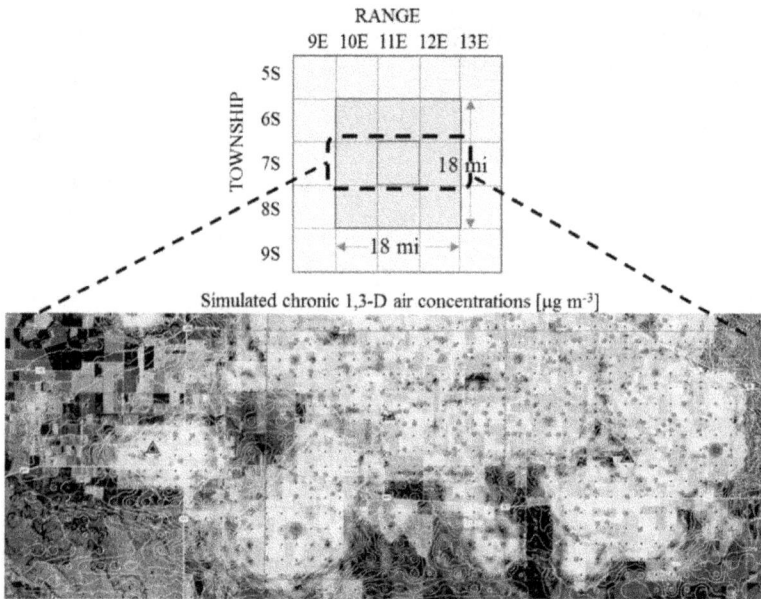

Figure 11.
Example of simulated SOFEA3 (i.e., AERMOD used) chronic air concentration predictions over a three-township area of Merced CA where field location, application rate and date are known and air concentrations at a central location in each township was monitored over a 14-month period.

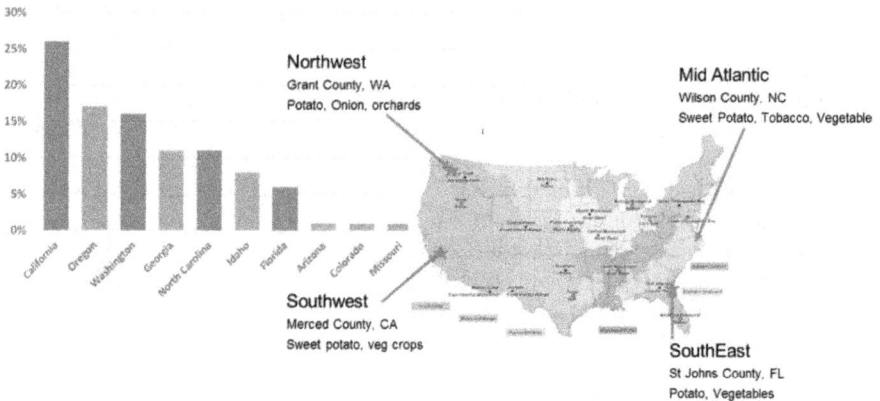

Figure 12.
Geographic region for SOFEA4 simulations for high 1,3-D use areas of the United States [11]. Results for the Southwest are found elsewhere [21].

from representative agronomic practices and large air sheds. SOFEA3/4 with AERMOD improved model predictions over what SOFEA2 with ISCST3 produced, due to the more realistic mixing height (MH) calculated by the weather pre-processor that subsequently resulted in higher concentrations during calm (stable air) periods. Advantages of using SOFEA4 to model fumigant concentrations over monitoring approaches for fumigant concentrations in air, include the ability to predict concentrations at a much greater temporal frequency (and spatial locations) than could be accomplished by monitoring alone. 1,3-D application data obtained from growers along with local weather data was used to parameterize SOFEA4, and it was found the Human Equivalent Concentrations

(HECs) for acute, short-term, subchronic, and chronic exposure for 1,3-D were
not exceeded for four study areas with intense 1,3-D use [11, 21].

4. Discussion

The SOil Fumigant Exposure Assessment (SOFEA) model was originally
developed in 2005 [3], expanded/refined multiple times over the last 14 years to
explore volatile pesticide exposure and bystander risk (most recent version being
SOFEA4 where C++ replaces VBA, and now with AERMOD the principle model
used). Multiple publications using SOFEA have been documented since SOFEA was
first developed [3, 11, 14, 17, 21, 31], with the most recent manuscript on SOFEA4
use for high use regions in the United States currently undergoing a journal review
process for publication [11]. SOFEA assembles sources (agricultural fields), various
management practices, source strengths (pesticide flux rates), weather data from
the region of interest, and executes an air dispersion model [AERMOD, ISCST3
(historical)] to simulate pesticide concentrations at user defined receptors whose
concentration predictions can be used in exposure and risk assessment procedures.
This book chapter describes the historical development of SOFEA up to the latest version (SOFEA4) including all attributes that have been described over the
years. SOFEA now uses AERMOD, the officially sanctioned USEPA regulatory air
dispersion model, in lieu of ISCST3 for air dispersion simulations. Recent SOFEA
simulation results were compared to the ambient air monitoring data collected
in an intensive 1,3-D fumigation field trial in Merced, CA specifically designed
to validate SOFEA against monitoring information. SOFEA4 (using AERMOD)
was shown to improve the prediction of high concentrations (and thus the annual
average concentration) compared to SOFEA2 (using ISCST3) and earlier versions of SOFEA. Better comparison against field observations using SOFEA4 was
attributed to the improved characterization of the Planetary Boundary Layer (PBL)
during calm period conditions (low wind), and more realistic Mixing Height (MH)
calculations that are employed by AERMOD compared to ISCST3 [11, 21].

The validated SOFEA4 model was further used to simulate 1,3-D air concentrations in three study areas across the US having significant 1,3-D use [11]. These
areas include Quincy, WA (representing the Pacific Northwest), Wilson, NC
(representing the Atlantic coastal plain), and St. John's, FL (representing GA/
FL). Including Merced, CA results [21], these four agricultural regions represent
land areas that account for ~95% of the 1,3-D sold in the USA, **Figure 12**. The most
recent publication (under review) of SOFEA deals with actual field locations and
1,3-D application parameters used for two annual product use cycles (2015–2016,
and 2016–2017) in multiple high use regions in the United States (as documented by
local growers in each study area [11]).

SOFEA has exceptional attributes and functionality compared to other similar
modeling tools for addressing the exposure and risk from the use of volatile (or
semi-volatile) pesticides. In a collaborative effort between Corteva Agriscience and
Exponent, Inc. (an Engineering and Scientific Consulting company), SOFEA3 was
upgraded to modern software engineering standards (renamed SOFEA4), and a new
graphical interface was developed with C++ and Qt to provide users an Integrated
Development Environment-like (IDE) experience in creating new simulation projects.
Summary statistics, moving averages and quantiles can be calculated efficiently over
millions of data points, comprising hourly concentrations over several years and thou-
sands of receptors. A beta version of SOFEA4 is currently being tested and evaluated
for release, and will be made publicly available online via Exponent, Inc. in mid-2019
for use with other volatile or semi-volatile chemicals for large-scale environmental and

risk assessment procedures. Parties interested in using SOFEA4 can contact Exponent Inc. directly to obtain both the user's guide and the latest version of the model.

5. Conclusions

SOFEA is used to predict soil fumigant and semi-volatile pesticide air concentrations under actual or projected use. SOFEA, first released in 2005 [3], has over a decade of development and refinement and is a comprehensive numerical tool that has been validated against many field trials and monitoring studies using 1,3-dichloropropene as summarized in this manuscript. Both the timing and magnitudes for more than 450 1,3-D treated fields in a 5 × 5 township domain were followed over a 1.5-year period [11, 21] in a field trial in CA specifically designed for SOFEA validation.

Examples of a soil physics model, CHAIN_2D, for flux predictions has also been used for source strength predictions in SOFEA, in addition to using field observations that are scaled by depth of soil incorporation and the time of the year when the application is made. Comparison of SOFEA predictions against other field studies has been good. SOFEA is a powerful tool to simulate air concentrations for regional agronomic conditions when multiple fields are simulated under typical agronomic conditions. SOFEA is especially useful for regulatory risk managers and product stewards who often are required to make decisions when only limited or incomplete data is available.

Chemical exposure information generated by SOFEA can be and has been used in a formalized risk assessment where risk to human populations is addressed. Understanding how various agronomic BMPs affect acute, sub chronic, and chronic exposure is an essential requirement for proper stewardship of volatile and semi-volatile pesticides, and SOFEA can be used for validation purposes or when limited, or no experimental evidence is available. SOFEA4 is being released to the public domain later in 2019 such that any user wanting to simulate air concentrations for volatile and semi-volatile chemicals in large and diverse airsheds can be incorporated, in support of (or in lieu of) monitoring trials.

List of abbreviations

1,3-D	1,3-dichloropropene
AERMOD	USEPA Regulatory Model (Gaussian plume)
ArcGIS	GIS for geographic information by ESRI
ArcView	Entry level of ArcGIS Desktop, a GIS software by ESRI
BMP	best management practice
CDPR	California Department of Pesticide Regulations
CHAIN_2D	computer program for 2-D variability saturated water flow, heat, solute transport
C++	general purpose program language
Crystal Ball™	software add-on for MS Excel developed by Decisioneering, Inc.
ESRI	Environmental Systems Research Institute
Exponent Inc.	US-based consulting company
GIS	geographic information system
Hydrus	water, heat, solute transport in saturated porous media (modeling software)
ISCST3	USEPA Industrial Source Complex Model (Gaussian plume)
PDF	probability density function

PLSS	public land survey system
PRZM3	USEPA Pesticide Root Zone Model
Qt	cross-platform framework for developing application software with a graphical user interface
SAP	USEPA Scientific Advisory Panel
SOFEA	SOil Fumigant Exposure Assessment system
TIF	totally impermeable film
USEPA	United States Environmental Protection Agency
VBA	visual basic for applications

Author details

Steven Cryer* and Ian van Wesenbeeck
Corteva Agriscience, USA

*Address all correspondence to: sacryer@dow.com

IntechOpen

References

[1] Cryer SA, van Wesenbeeck IJ. Predicted 1,3-dichloropropene air concentrations resulting from tree and vine applications in California. Journal of Environmental Quality. 2001;**20**:1887-1895

[2] Cryer SA, van Wesenbeeck IJ, Knuteson JA. Predicting regional emissions and near-field air concentrations of soil fumigants using modest numerical algorithms: A case study using 1,3-dichloropropene. Journal of Agricultural and Food Chemistry. 2003;**51**:3401-3409

[3] Cryer SA. Predicting soil fumigant acute, sub-chronic, and chronic air concentrations under diverse agronomic practices. Journal of Environmental Quality. 2005;**34**:2197-2207

[4] USEPA. Industrial Source Complex (ISC3) Dispersion Model User's Guide; Volumes I and II. EPA-454/B-95-003a and b. Research Triangle Park, North Carolina: U.S. Environmental Protection Agency; 1995

[5] AERMOD, United States Environmental Protection Agency. AERMOD: Description of Model Formulation. EPA-454/R-03-004. Research Triable Park, North Carolina: U.S. Environmental Protection Agency; 2004. Available from: http://www.epa.gov/scram001/7thconf/aermod/aermod_mfd.pdf

[6] Cimorelli AJ, Perry SG, Venkatram A, Weil JC, Paine RJ, Wilson RB, et al. AERMOD: Description of Model Formulations. EPR-454/R-03-004. Research Triangle Park, North Carolina: U.S. Enviromental Protection Agency; 2004

[7] Hao J, He D, Wu Y, Fu L, He K. A study of the emission and concentration distribution of vehicular pollutants in the urban area of Beijing. Atmospheric Environment. 1999;**34**:454-465

[8] Sullivan DA, Holdworth MT, Hinka DJ. A Monte Carlo-based dispersion modeling of off-gassing releases from the fumigant metam-sodium for determining distances to exposure endpoints. Atmospheric Environment. 2004;**38**:2741-2481

[9] Heckel PF, LeMasters GK. The use of AERMOD air pollution dispersion models to estimate residual ambient concentrations of elemental mercury. Water, Air, and Soil Pollution. 2011;**219**(1-4):377-388

[10] Jittra N, Pinthong N, Thepanondh S. Performance evaluation of AERMOD and CALPUFF air dispersion models in industrial complex area. Air, Soil and Water Research. 2015;**8**:87-95. DOI: 10.4137/ASWR.S32781

[11] de Cirugeda Helle O, van Wesenbeeck IJ, Cryer SA. SOFEA modeling of 1,3-Dichloropropene concentrations in ambient air in high fumigant use areas of the United States. American Chemical Society (ACS). 255th National meeting. Boston, MA. August 19-23, 2018

[12] Yu C, Zheng C. HYDRUS: Software for flow and transport modeling in variably saturated media. Software Spotlight, Ground Water. 2010;**48**(6):787-791

[13] Šimůnek J, van Genuchten MTh. The CHAIN_2D Code for Simulating Two-Dimensional Movement of Water, Heat, and Multiple Solutes in Variably-Saturated Porous Media, Version 1.1. Research Report No. 136; Riverside, California: U.S. Salinity Laboratory, USDA, ARS; 1994

[14] USEPA Agency Scientific Advisory Panel 2004. Fumigant Bystander Exposure Model Review: Soil Fumigant Exposure Assessment System (SOFEA) Using Telone as a Case Study. Available

from: https://archive.epa.gov/scipoly/sap/meetings/web/html/090904_mtg.html [Accessed: 10 September 2004]

[15] Ross LJ, Johnson B, Kim KD, Hsu J. Prediction of methyl bromide flux from area sources using the ISCST model. Journal of Environmental Quality. 1996;**25**:885-891

[16] Yates SR, Ashworth DJ, Zheng W, Knuteson J, van Wesenbeeck IJ. Effect of deep injection on field-scale emissions of 1,3-dichloropropene and chloropicrin from bare soil. Atmospheric Environment. 2016;**137**:135-145

[17] Cryer SA, van Wesenbeeck IJ. Coupling field observations, soil modeling, and air dispersion algorithms to estimate 1,3-dichloropropene and chloropicrin flux and exposure. Journal of Environmental Quality. 2011;**2011**(40):1450-1461. (Invited Paper for Special Submissions: Agricultural Air Quality)

[18] Rubinstein RY. Simulation and the Monte Carlo Method. New York, NY, USA: John Wiley & Sons, Inc.; 1981. ISBN: 0471089176

[19] Yakowitz S. Computational Probability and Simulation. Reading, MA: Addison-Wesley Publishers; 1977

[20] USEPA. Guiding Principles for Monte Carlo Analysis. EPA/630/R-97/001. Risk Assessment Forum. Washington, DC: U.S. Environmental Protection Agency; 1997

[21] van Wesenbeeck IJ, Cryer SA, de Cirugeda Helle O, Li C, Driver J. Comparison of regional air dispersion simulation and ambient air monitoring data for the soil fumigant 1,3-dichloropropene. Science of the Total Environment. 2016:569-570 603-610

[22] USEPA. AERMOD Model Formulation and Evaluation. Office of Air Quality Planning and Standards, Air Quality Assessment Division, Research Triangle Park, NC: EPA-454/R-18-003 U.S. Environmental Protection Agency. 2018. Available from: https://www3.epa.gov/ttn/scram/models/aermod/aermod_mfed.pdf

[23] Šimůnek J, Van Genuchten MTh, Sejna M. The HYDRUS software package for simulating two- and three-dimensional movement of water, heat and multiple solutes in variable-saturated media. Technical manual, version 3.0, PC-Progress, Prague, Czech Republic, 2018. p. 274

[24] van Genuchten MT, Šimůnek J, Leij FL, Toride N, Šejna M. STANMOD: Model use, calibration and validation, special issue standard/engineering procedures for model calibration and validation. Transactions of the ASABE. 2012;**5**(4):1353-1366

[25] PRZM-3. A model for Predicting Pesticide and Nitrogen Fate in the Crop Root and Unsaturated Soil Zones: User's Manual for Release 3.12.2. EPA/600/R-05/111. Ecosystems Research Division, National Exposure Research Laboratory, Athens, GA, 30605-2700, U.S. Environmental Protection Agency, Office of Research and Development, Washington, DC; 2005

[26] Cryer SA, van Wesenbeeck IJ. Estimating field volatility of Soil Fumigants Using CHAIN_2D: Mitigation methods and comparison against chloropicrin and 1,3-dichloropropene field observations. Environmental Modeling and Assessment. 2010;**15**:309-318

[27] Wang J, Huang G, Zhan H, Mohanty BP, Zheng J, Huang Q, et al. Evaluation of soil water dynamics and crop yield under furrow irrigation with a two-dimensional flow and crop growth coupled model. Agricultural Water Management. 2014;**141**(10-22):2014. DOI: 10.1016/j.agwat.2014.04.007

[28] Abdou HM, Flury M. Simulation
of water flow and solute transport
in free-drainage lysimeters and field
soils with heterogeneous structures.
European Journal of Soil Science.
2004;**55**(2):229-241

[29] Spurlock F, Johnson B, Tuli A,
Gao S, Tao J, Sartori F, et al. Simulation
of fumigant transport and volatilization
from tarped broadcast applications.
Vadose Zone Journal. 2013;**12**(3):1-10

[30] Scire JS, Strimaitis DG, Yamartino RJ.
A User's Guide for The CALPUFF
Dispersion Model. Concord, MA: Earth
Tech., Inc.; 2002

[31] van Wesenbeeck IJ, Cryer SA,
Havens PL, Houtman BA. Use of SOFEA
to predict 1,3-D concentrations in
air in high use regions of California.
Journal of Environmental Quality.
2011;**40**:1462-1469

Chapter 4

Atmospheric Air Pollution in Nigeria: A Correlation between Vehicular Traffic and Criteria Pollutant Levels

Yahaya Abbas Aliyu, Joel Ondego Botai,
Aliyu Zailani Abubakar, Terwase Tosin Youngu,
Jimoh Olanrewaju Sule, Mohammed Wachin Shebe
and Mohammed Ahmed Bichi

Abstract

In Nigeria, the rising levels of used/poorly maintained vehicles are contributing to most urban air pollution with possible repercussion on the general public health. This study evaluates the inferences of vehicular traffic surge on outdoor pollutant measurement using Zaria, northern Nigeria, as a case study. The study collected a 1-year time-series dataset for the vehicular count and the respective outdoor criteria pollutant measurements over 19 study sites. The vehicular traffic was categorized into motorcycles (2-W), tricycles (3-W), cars, buses, light-duty vehicles (LDV) and heavy-duty vehicles (HDV). The outdoor pollutants that were measured include carbon monoxide (CO), sulfur dioxide (SO_2) and particulate matter ($PM_{2.5}/PM_{10}$). We utilized validated portable monitors (CW-HAT200 particulate counter and the MSA Altair 5x multigas sensor) for the outdoor measurements during December 2015–November 2016. The observed measurements for the validation procedure were normally distributed [kurtosis (0.301); skewness (−0.334)] and coefficient of determination (R2 ≥ 0.808). The time-series analysis of particulate matter (PM) measurements displayed alarming concentrations levels. Combined vehicular traffic density analysis revealed significant contribution (R ≥ 0.619) to the population exposed outdoor pollutant measurements. The 2-W (motorcycle) was found to be the vehicular category that attributed the most significant relationship with observed outdoor pollutant measurements.

Keywords: urban air quality, vehicular traffic, portable sensors, criteria pollutants, Zaria-Nigeria

1. Introduction

In most developing countries, atmospheric pollution continues to affect exposed population health [1–3]. In Africa, air quality studies are devising alternative and reliable means to obtain pollutant measurements for research. The approach

$\mu g\,m^{-3}$	Microgram per meter cube
2-W	Two-wheeler (motorcycle)
3-W	Three-wheeler (tricycle)
CO	Carbon monoxide
HDV	Heavy-duty vehicle
LDV	Light-duty vehicle
$PM_{2.5}$	Particulate matter, with a diameter of <2.5 μm
PM_{10}	Particulate matter, with a diameter of <10 μm
ppm	Parts per million
SO_2	Sulfur dioxide
TSP	Totally suspended particles

Table 1.
List of abbreviations and units.

includes reliable validation of sampling techniques that contribute to the up-to-date understanding of criteria pollutants, maintenance outflow and technical know-how [4].

Nigeria's rising population is escalating anthropogenic activities within its territory without any reliable information on its air quality [5]. The atmospheric air quality of most of its urban cities continues to remain exposed to the growing, poorly managed vehicular traffic from ineffective fuel combustion [6]. The situation is familiar, however, the motivation to address it lingers ambiguously.

The rising levels of used/poorly managed vehicular operations remains an unnoticed contributor to urban atmospheric air pollution. While the literature has established alarming pollutant levels and how they contribute to the respiratory wellbeing of the exposed population [3], there is a need to further establish the relationship between the categories of the existing vehicular traffic surge and corresponding criteria pollutant levels observed. This will facilitate the process of the traffic-related atmospheric air pollution management plan in many Nigerian cities. For familiarity with the terminology, **Table 1** highlights a list of abbreviations and units utilized for this study.

2. Methodology

2.1 Study area

Zaria metropolis described in **Figure 1**, has an estimated area of 296.04 km². Its estimated population as reported in 2014 is 938,521. The city climate characteristics are divided into two. The dry season ranges from October to May, and the rainy season ranges from June to September. The altitude is averagely 670 m above mean sea level [7]. Major road intersections are the concept adopted for the selection of the 19 study sites.

2.2 Instrumentation and methods

There is increasing use of portable devices for examining outdoor air (atmospheric) quality. With comparison to established reference devices, their reliability allows for effective real-time data acquisition, especially in limited resource

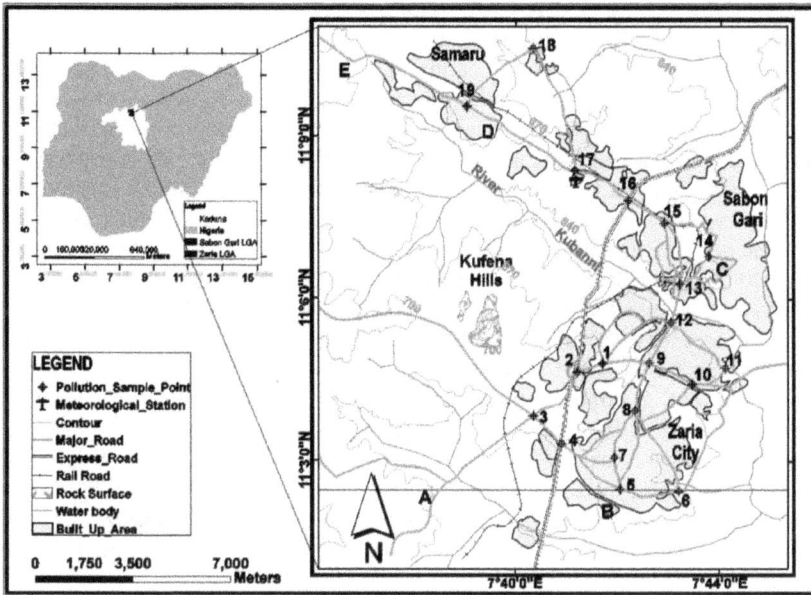

Figure 1.
The 19 study sites adopted for study data acquisition.

environments [8]. This study employed the CW-HAT200 particulate counter and the MSA Altair 5x multi-gas sensor to collect particulate matter ($PM_{2.5}$ and PM_{10}) while the MSA Altair 5x collect carbon monoxide (CO) and sulfur dioxide (SO_2) respectively. The instrument re-calibration was conducted using manufacturer's span calibration mixed gas specifications.

Owing to the unavailability of real-time reference air pollution monitors within the study region, the devices were validated the portable pollutants monitors using the WHO air filter sampling model Eq. (1). To validate the portable devices, total suspended particulates (TSP) were collected at two distinct sample test stations at 1.5 m above the existing ground level. Validation site 1 had dense outdoor traffic activity, while validation site 2 had minimal outdoor traffic activity tagged control site. The validation samples and synchronized portable monitor measurements were obtained across three epochs, that are, morning, afternoon and evening for 17 days. TSP is described as particulate fraction ranging from 0.1 to about 100 μm in size (diameters). Particulates matter $PM_{2.5}$ (diameter < 2.5 μm) and PM_{10} (diameter < 10 μm) fall within the specified range. Based on [9] which identified a significant relationship between total suspended particulates, PM_{10} and $PM_{2.5}$ and [10] which reported that there is a significant correlation among pollutant emissions resulting from a common source, the study validated the portable devices using the WHO air sampling filter technique. Eq. (1) describes the WHO air sample model technique [11].

$$total\ suspended\ particulates\ (\mu g\ m^{-3}) = \frac{M_S - M_O}{V} \qquad (1)$$

where M_O is the filter paper mass without TSP samples, M_S is the filter paper mass with TSP samples, V is the TSP volume. To determine the concentration ($\mu g\ m^{-3}$), model Eq. (1) was divided by the sample time (in hours).

In line with Eq. (1), the validation samples were collected individually on filter papers and collocating pollutant measurements with the portable device over the study duration. The particulate filter samples were processed in the laboratory to obtain their individual concentrations using Eq. (1). They were then compared with the separately recorded collocating pollutant measurements from the portable devices. The collocating measurements were then analyzed using linear regression and bias, for the validation of the portable monitors. The analysis is described in **Figures 2** and **3**. The observed measurements for the validation procedure were normal distributed [skewness (-0.334); kurtosis (0.301)]. The study adopted two performance indicators for the purpose of validating the portable pollutant instrument. The performance indicators are The Bland-Altman agreement plot and the coefficient of determination (R^2). The Bland-Altman plot evaluates the systematic bias between the two measurements techniques, while the coefficient of determination indicates how strongly related the pair(s) of variables are. The Bland-Altman agreement plot can be seen in **Figure 2**.

From **Figure 2**, it can be seen that there is no significant systematic difference in the measurements. Additionally, the coefficient of determination (R^2) across the two test sites showed that the TSP measurements from the WHO model technique and criteria pollutant measurements from the MSA Altair 5x/CW-HAT200 devices were significantly correlated. The linear regression can be seen in **Figure 3**. **Figures 2** and **3** illustrate that the reliability of the portable pollutant monitors has been validated based on [9, 10].

(a)   (b)

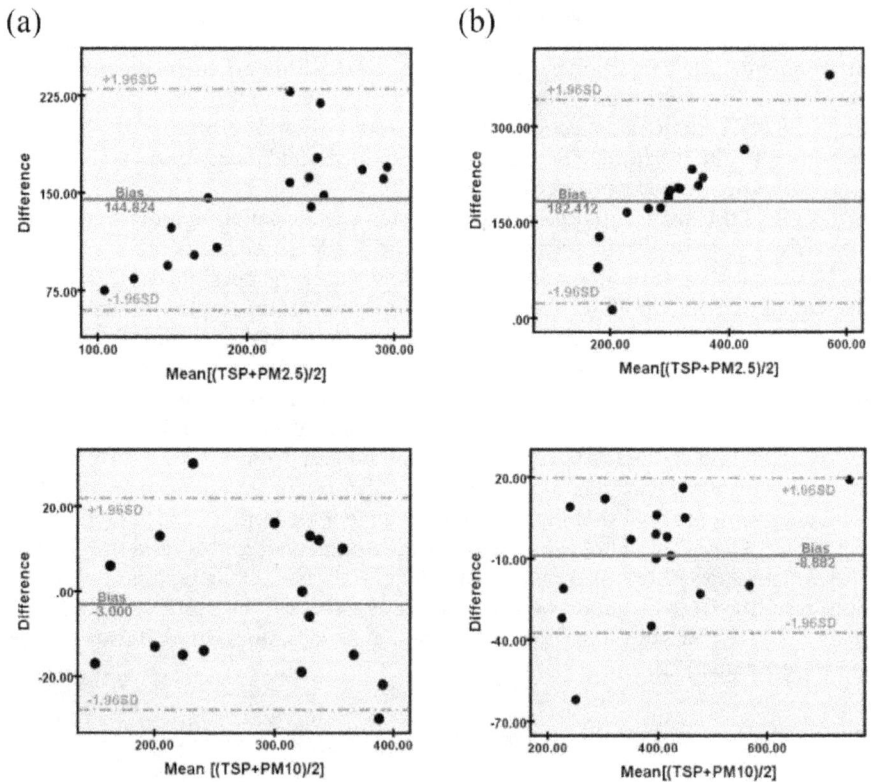

Figure 2.
Bland-Altman bias plot highlighting the agreement of observed validation measurements ($PM_{2.5}$ and PM_{10}) within the 95% confidence interval: (a) less densely populated site and (b) densely populated site.

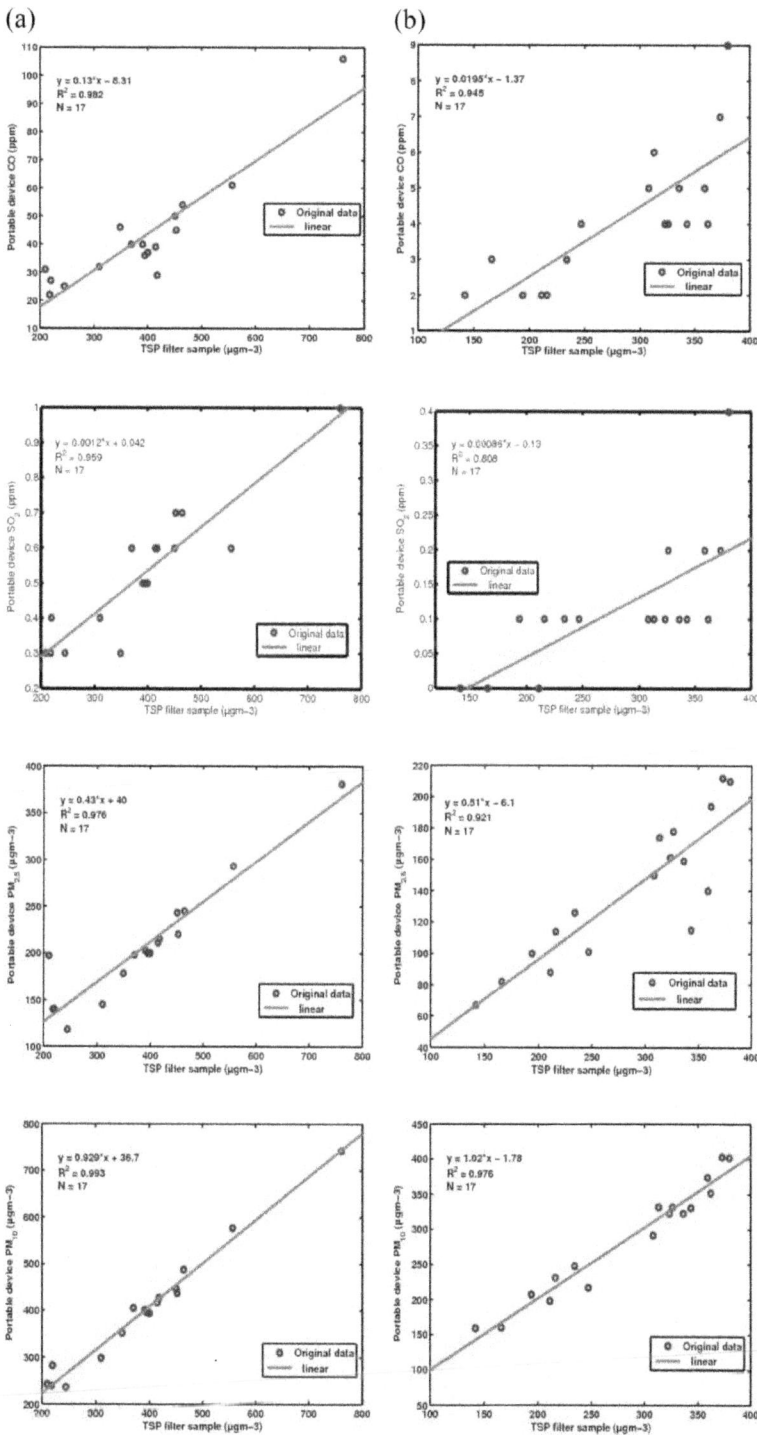

Figure 3.
Scatter plots showing the linear regression and coefficient of determination between the TSP and the portable monitor samples: (a) densely populated site and (b) control site.

Study sites	s1	s2	s3	s4	s5	s6	s7	s8	s9	s10	s11	s12	s13	s14	s15	s16	s17	s18	s19
s1	1	0.997	0.878	0.998	0.983	0.858	0.997	0.999	0.995	0.997	0.986	0.994	0.996	0.988	0.994	0.995	0.989	0.960	0.995
s2		1	0.907	0.993	0.992	0.888	0.989	0.998	0.985	1.000	0.993	0.984	0.987	0.974	0.988	0.985	0.994	0.977	0.997
s3			1	0.851	0.951	0.997	0.841	0.895	0.834	0.910	0.906	0.837	0.847	0.797	0.833	0.826	0.897	0.961	0.897
s4				1	0.971	0.828	0.998	0.994	0.995	0.992	0.983	0.993	0.994	0.993	0.999	0.998	0.989	0.950	0.993
s5					1	0.938	0.967	0.989	0.963	0.993	0.982	0.964	0.969	0.944	0.961	0.959	0.982	0.988	0.986
s6						1	0.820	0.878	0.815	0.892	0.881	0.820	0.829	0.775	0.808	0.803	0.873	0.944	0.875
s7							1	0.994	0.999	0.989	0.975	0.998	0.998	0.997	0.996	0.999	0.980	0.937	0.987
s8								1	0.991	0.999	0.985	0.992	0.993	0.982	0.990	0.990	0.990	0.968	0.996
s9									1	0.985	0.967	0.999	1.000	0.997	0.992	0.998	0.973	0.928	0.982
s10										1	0.991	0.985	0.988	0.974	0.987	0.985	0.993	0.978	0.997
s11											1	0.963	0.968	0.957	0.980	0.971	0.995	0.981	0.992
s12												1	0.999	0.996	0.990	0.997	0.971	0.928	0.981
s13													1	0.994	0.990	0.996	0.973	0.933	0.983
s14														1	0.993	0.998	0.965	0.909	0.974
s15															1	0.997	0.988	0.943	0.991
s16																1	0.979	0.931	0.985
s17																	1	0.981	0.978
s18																		1	0.998
s19																			1

Table 2.
Pearson's correlation coefficient matrix of seasonal pollutant measurement across the 19 study sites (significant at 0.01 levels).

With the above-described validation, the portable instruments were utilized to commence the measurement of ground level roadside pollution concentrations. The duration of the sampling measurement was from 01 December 2015 to 30 November 2016. The outdoor concentration levels were observed using the approach described in [12, 13]. The vehicular traffic count was also conducted to obtain the volume of vehicles contributing to the outdoor air pollution across the sampling sites. The vehicular count was obtained to determine the contributory level of vehicular density to outdoor air pollution. The vehicles are categorized as follows: motorcycles (2-W), tricycles (3-W), cars, buses, light-duty vehicles (LDV) and heavy-duty vehicles (HDV). The study analysis was performed using software: SPSS, Microsoft Excel and MATLAB.

3. Results and discussion

Table 2 highlights the dispersal relationship of the observed CO, SO_2, $PM_{2.5}$ and PM_{10} across the 19 study sites. This was achieved using Pearson's correlation coefficient. The inter-study-site correlation matrix (**Table 2**), showed that the relationship of the measured pollutants was significant at the 0.01 level across all the study sites. And only study site 6 (a control site) revealed lower coefficient values in comparison to the remaining study sites. From **Table 2**, study sites 2 and 9 produced a perfect relationship with site 10 and site 13, respectively.

Study site	2-W	3-W	Car	Bus	LDV	HDV
1	16,034 ± 17	3186 ± 4	10,242 ± 11	5613 ± 6	958 ± 1	1417 ± 2
2	15,111 ± 16	2955 ± 4	8443 ± 9	4971 ± 6	643 ± 1	1204 ± 2
3	8554 ± 8	888 ± 1	3021 ± 3	1177 ± 1	2571 ± 1	641 ± 1
4	19,948 ± 22	3731 ± 4	11,785 ± 12	7279 ± 8	1561 ± 2	3444 ± 4
5	11,688 ± 12	2063 ± 2	2960 ± 4	1418 ± 2	340 ± 1	571 ± 1
6	5602 ± 6	615 ± 1	1585 ± 2	542 ± 1	241 ± 1	412 ± 1
7	18,012 ± 18	3954 ± 5	4045 ± 5	6656 ± 7	502 ± 2	442 ± 1
8	17,069 ± 17	3556 ± 4	5153 ± 5	6353 ± 7	428 ± 1	211 ± 1
9	27,008 ± 27	5529 ± 7	16,307 ± 17	9352 ± 10	1495 ± 2	1628 ± 2
10	14,870 ± 15	3575 ± 4	8628 ± 9	3667 ± 5	784 ± 2	1320 ± 1
11	14,453 ± 16	4446 ± 6	7089 ± 8	2321 ± 3	296 ± 1	369 ± 1
12	27,058 ± 28	5720 ± 6	15,746 ± 17	9643 ± 10	1436 ± 2	929 ± 2
13	22,012 ± 23	4982 ± 6	9559 ± 10	8551 ± 9	919 ± 2	537 ± 1
14	28,897 ± 29	6205 ± 7	5123 ± 6	6797 ± 8	897 ± 2	500 ± 1
15	17,482 ± 18	3736 ± 5	11,748 ± 13	6761 ± 7	1343 ± 2	899 ± 2
16	20,678 ± 22	4672 ± 6	22,656 ± 24	13,050 ± 14	2405 ± 3	2666 ± 4
17	11,167 ± 12	2241 + 3	12,647 ± 13	6447 ± 8	1013 ± 2	1131 ± 2
18	6710 ± 7	756 ± 1	4194 ± 5	2528 ± 3	364 ± 1	713 ± 1
19	10,529 ± 11	2048 ± 3	9781 ± 10	6063 ± 7	872 ± 2	896 ± 2
Total	312,882	64,858	170,712	109,189	19,068	19,930

Table 3.
Vehicular traffic density (total ± average per 3 min) across the 19 sampling sites in the study.

The traffic count for the individual sampling site per epoch was computed based on the vehicular category, as shown in **Table 3**. In general, the study site with the highest weighted average of the criteria pollutants measured over the 19 study locations is study site 14. The reason for the high measurements is because the site is within the study area's main market (Sabon-Gari market) with the highest average count of 2-W and 3-W vehicle density (**Table 3**). The traffic volume was determined by direct counting the traffic during the daily sampling epoch for the study period (1 year).

Figure 4 displays the time-series plots of vehicular traffic count and resulting criteria pollutants measurements collected over selected study sites 3, 9 and 15. It can be observed that the study sites 3 which is a control site, did have the majority of its pollutant concentration levels below 40 ppm, 0.6 ppm, 300 µg m^{-3} and

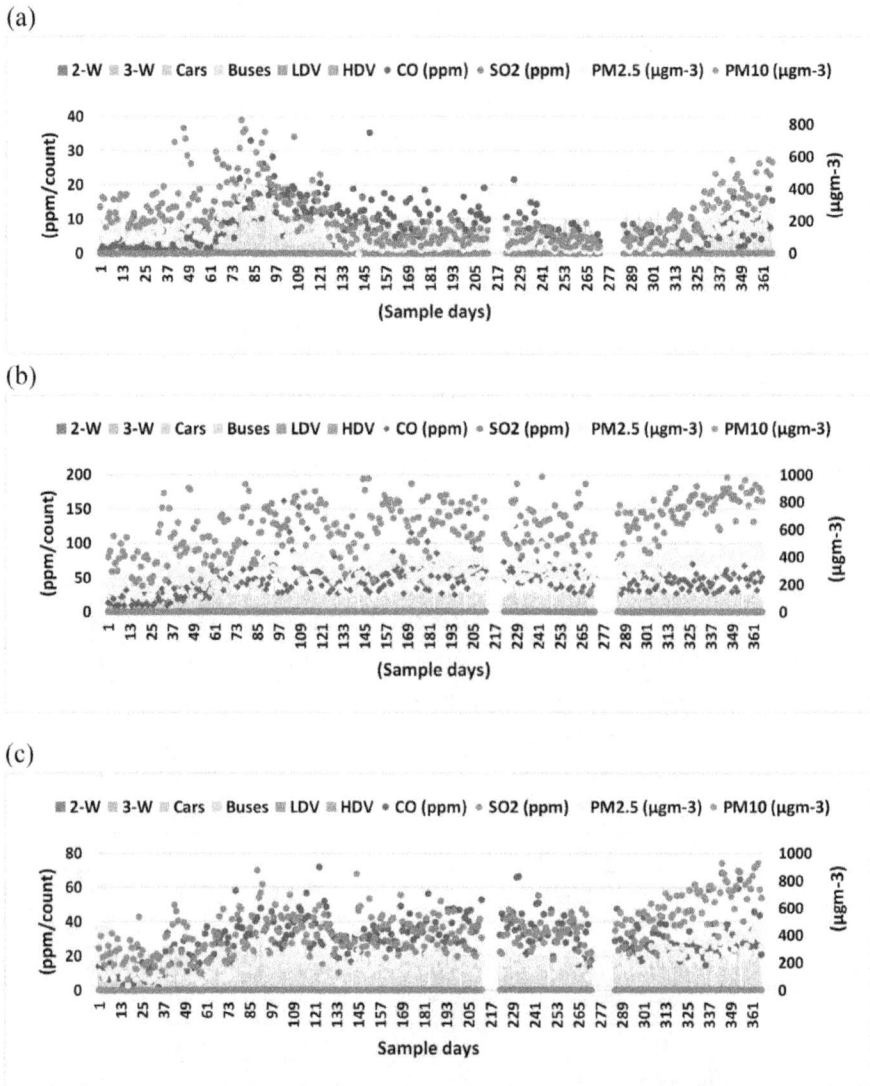

(a)

(b)

(c)

Figure 4.
Time-series of the weighted average for the vehicular traffic count against the measured criteria pollutants over randomly selected sites 3, 9, and 15 for the 366 days duration: (a) Study site 3, (b) Study site 9, and (c) Study site 15.

600 µg m^{-3} for CO, SO$_2$, PM$_{2.5}$ and PM$_{10}$ respectively. Except for PM during the Harmattan season which falls between sample days 1–91. For sites 9 and 15, the majority of the observed criteria pollutant measurements were above the earlier described values.

Table 4 presents the computed 1-year weighted average of the measured criteria pollutants concentrations across the 19 study sites [14]. Study sites 3 and 6 recorded the least pollutant measurements CO/SO$_2$ and PM$_{2.5}$/PM$_{10}$. This could attribute to minimal population activities at the sites. The sites 3, 6 and 18 were actually selected to serve as control sites for the study. From Table 4, the weighted average of the observed criteria pollutants for the study area is deduced as CO (29.220 ppm), SO$_2$ (0.319 ppm), PM$_{2.5}$ (219.729 µg m^{-3}) and PM$_{10}$ (451.958 µg m^{-3}).

Additionally, the weighted average computed for the observed criteria pollutant was compared against the stipulated guidelines in the WHO air quality document [15]. The comparison revealed that the weighted average of criteria pollutants observed over the 19 study sites did exceed the WHO stipulated threshold (blue line across bar charts) for SO$_2$, PM$_{2.5}$ and PM$_{10}$ in all the study sites, except for CO, whose weighted average stayed within the stipulated limits only in sites 3, 6 and 18. This is illustrated in Figure 5.

Pearson's correlation matrix was utilized to investigate the seasonal level of association between measured criteria pollutants and traffic activities within the 19

Site	Latitude	Longitude	Description	CO (ppm)	SO$_2$ (ppm)	PM$_{2.5}$ (µg m^{-3})	PM$_{10}$ (µg m^{-3})
1	11.080	7.695	Kofar Kibo	33.036	0.363	258.873	528.000
2	11.078	7.686	Danmagaji, Wusasa	20.838	0.264	214.720	432.571
3	11.064	7.673	Madaci, Saye	7.994	0.159	117.177	232.246
4	11.054	7.682	Gwargwaje	29.703	0.351	250.294	509.957
5	11.044	7.701	Kofar Gayan	16.811	0.212	182.562	372.982
6	11.041	7.720	Kofar Kona	4.586	0.137	99.068	202.008
7	11.051	7.699	Zaria City market	38.281	0.383	276.448	561.482
8	11.066	7.706	Babban Dodo	27.242	0.290	220.292	448.332
9	11.081	7.710	Kofar Doka	46.844	0.449	312.469	631.429
10	11.074	7.725	Banzazzau	22.880	0.260	208.111	424.255
11	11.079	7.735	FCE/Ungwan Kaya	19.728	0.243	179.426	367.067
12	11.093	7.717	Agwaro, Tudun Wada	55.959	0.525	328.026	662.063
13	11.104	7.721	PZ	38.848	0.399	282.524	573.486
14	11.113	7.730	Sabon Gari market	65.073	0.627	342.588	704.262
15	11.124	7.715	MTD	29.600	0.302	173.255	448.810
16	11.130	7.703	Kwangila bridge	50.130	0.465	282.891	576.923
17	11.139	7.686	Aviation by NITT road	19.180	0.238	167.004	352.324
18	11.177	7.672	Basawa by Hayin Dogo	8.795	0.167	93.807	189.041
19	11.159	7.651	Samaru market	19.652	0.244	185.319	369.965

Table 4.
The 1-year weighted average of the observed pollutants (N = 19, 104).

(a)

(b)

(c)

(d)

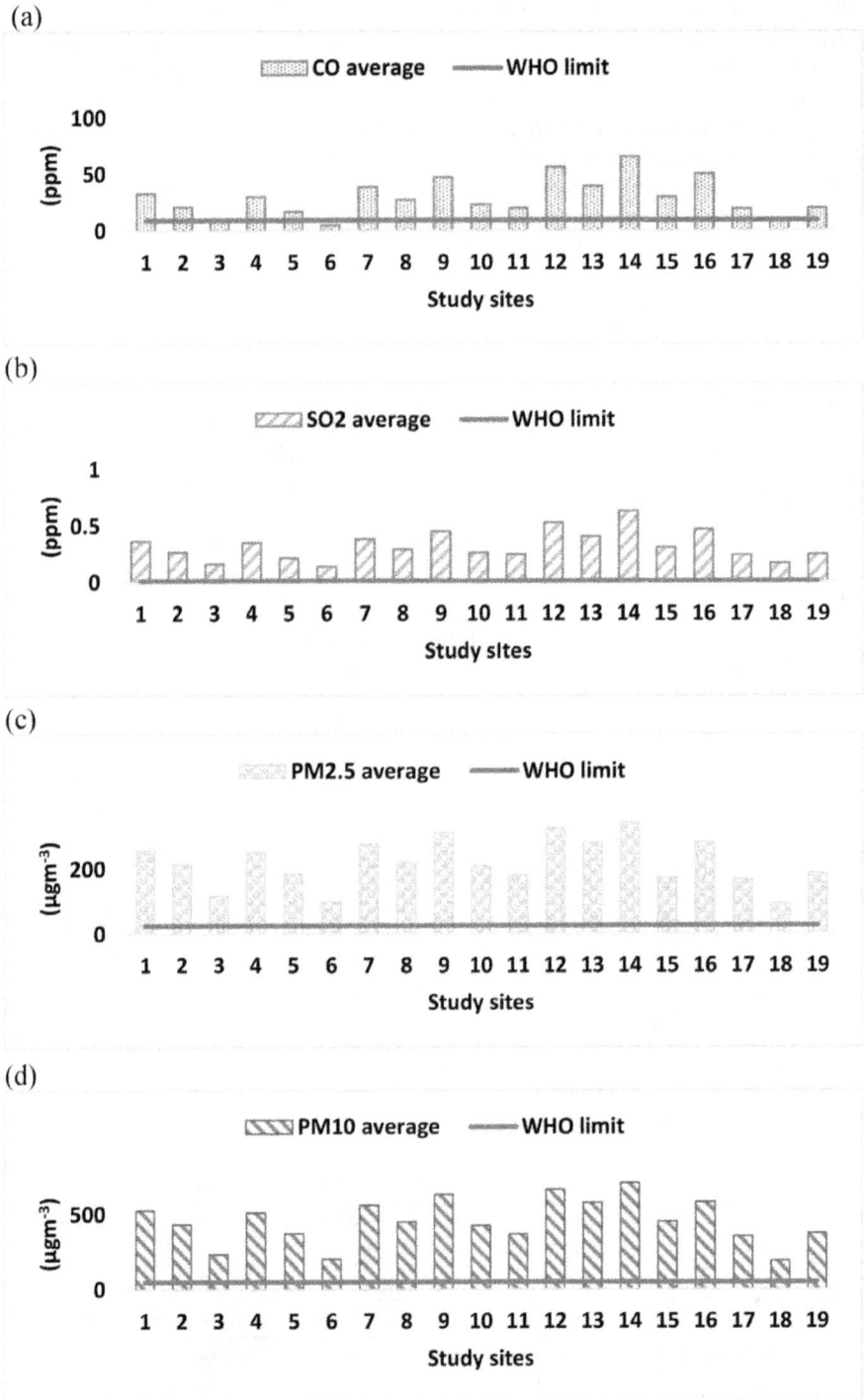

Figure 5.
The comparison of weighted criteria pollutants average: (a) CO; (b) SO₂; (c) PM₂.₅; and (d) PM₁₀ against the WHO air quality guidelines.

Pollutants	Seasons	CO				SO$_2$				PM$_{2.5}$				PM$_{10}$				Traffic count			
		DJF	MAM	JJA	SON	DJF	MAM	JJA	SON	DJF	MAM	JJA	SON	DJF	MAM	JJA	SON	DJF	MAM	JJA	SON
CO	DJF	1	0.983	0.975	0.940	0.962	0.957	0.963	0.927	0.778	0.951	0.941	0.900	0.779	0.948	0.941	0.910	0.870	0.876	0.837	0.790
	MAM		1	0.992	0.972	0.976	0.984	0.988	0.958	0.782	0.968	0.960	0.916	0.783	0.966	0.960	0.930	0.882	0.898	0.854	0.814
	JJA			1	0.984	0.969	0.976	0.995	0.975	0.780	0.970	0.971	0.934	0.781	0.969	0.971	0.949	0.882	0.898	0.861	0.822
	SON				1	0.946	0.949	0.979	0.992	0.749	0.949	0.958	0.943	0.752	0.951	0.959	0.954	0.894	0.903	0.882	0.860
SO$_2$	DJF					1	0.985	0.968	0.929	0.747	0.930	0.922	0.879	0.750	0.929	0.922	0.898	0.816	0.837	0.781	0.737
	MAM						1	0.983	0.930	0.751	0.935	0.923	0.861	0.752	0.931	0.923	0.885	0.828	0.842	0.791	0.749
	JJA							1	0.972	0.789	0.966	0.963	0.917	0.790	0.963	0.963	0.933	0.867	0.887	0.845	0.804
	SON								1	0.761	0.955	0.971	0.968	0.766	0.957	0.972	0.975	0.909	0.914	0.891	0.864
PM$_{2.5}$	DJF									1	0.871	0.837	0.778	0.999	0.864	0.836	0.785	0.673	0.691	0.664	0.619
	MAM										1	0.989	0.952	0.871	0.999	0.989	0.960	0.885	0.904	0.858	0.815
	JJA											1	0.977	0.837	0.992	1.000	0.984	0.892	0.911	0.874	0.831
	SON												1	0.783	0.960	0.977	0.996	0.919	0.922	0.896	0.861
PM$_{10}$	DJF													1	0.864	0.837	0.790	0.676	0.694	0.667	0.622
	MAM														1	0.992	0.967	0.886	0.905	0.861	0.819
	JJA															1	0.985	0.893	0.911	0.875	0.832
	SON																1	0.898	0.903	0.872	0.834
Traffic count	DJF																	1	0.985	0.973	0.955
	MAM																		1	0.986	0.964
	JJA																			1	0.992
	SON																				1

Table 5.
Seasonal correlation of the measured pollutants against the traffic variables (significant at 0.01 levels).

	2-W	3-W	Cars	Buses	LDV	HDV
CO	0.865*	0.793*	0.523	0.665*	0.542	0.433
SO$_2$	0.710*	0.694	0.422	0.587*	0.458	0.352
PM$_{2.5}$	0.763*	0.719*	0.465	0.593*	0.461	0.361
PM$_{10}$	0.766*	0.720*	0.468	0.600*	0.462	0.359

The gradient of the shaded cells highlights (in decreasing order) the ranking of correlation of the various categories of vehicles to the observed criteria pollutants.

Table 6.
Statistical correlation between vehicular categories collated at the study sites against the criteria pollutant measurements.

sampling locations. The data capture period was categorized into seasons that include December-January-February (DJF); March-April-May (MAM); June-July-August (JJA) and September-October-November (SON). This aims to appraise the environmental implication of road traffic movement to outdoor air pollution in Zaria across the seasons. From **Table 5**, it can be observed that all the measured variables were correlated positively at 0.01 p-levels. The analysis also indicates that the traffic activities (that is, the vehicular counts at the time of criteria pollutant observations) contributed significantly to observed criteria pollutants concentration levels except for the December-January-February (DJF) season. The DJF season (**Table 5**, red text) recorded lower correlation coefficients compared to the remaining seasons. The lower Pearson's coefficients during the DJF season can be attributed to the Harmattan and the holiday season within the study area. The Harmattan season is characterized by natural dusty-windy conditions and low temperatures, while the holiday season attributed to the lesser than usual traffic activities within the study area. From **Table 5**, this study concludes that emissions from vehicular activities are significantly responsible for measured pollutants observations in this study.

The contribution of traffic variables to the outdoor air pollution level is further evaluated with the consideration of the various vehicle categories (2-W, 3-W, cars, buses, LDV and HDV). **Table 6** described the contributory relationship between the observed criteria pollutants and the vehicular category. From **Table 6**, it can be observed that 2-W (motorcycles) counts showed the strongest relationship with the individual criteria pollutants measured, this followed by the 3-W (tricycles) and then buses. These findings confirmed the theory of the terrible state of these categories of the vehicle in the study.

4. Conclusions

Urban air quality management remains a continuous task for Nigerian policymakers. This study assessed the implication of varying categories of vehicular traffic on outdoor air pollution over a developing Nigeria city. This was achieved through day-time primary data capture of vehicular traffic and corresponding criteria pollutant measurements over a period of 1 year (December 2015–November 2016). The result of the criteria pollutant measurements was alarmingly high as confirmed by similar studies. Furthermore, the study concluded that the combined vehicular traffic did contribute significantly (R ≥ 0.619) to the observed pollutant measurements all through the study. The 2-W (motorcycle) was found to be the vehicular category that attributed the most significant relationship with observed outdoor pollutant measurements. This is followed by the 3-W (tricycles) and buses. The findings of the study will assist Nigerian policymakers on decisive steps for vehicular worthiness to urban air quality management.

Acknowledgements

This study is supported by a postgraduate bursary to the first author, from the University of Pretoria, South Africa and the Ahmadu Bello University, Zaria, Nigeria.

Conflict of interest

The authors declare no conflict of interest.

Author details

Yahaya Abbas Aliyu[1,3*], Joel Ondego Botai[1,2], Aliyu Zailani Abubakar[3], Terwase Tosin Youngu[3], Jimoh Olanrewaju Sule[3], Mohammed Wachin Shebe[3] and Mohammed Ahmed Bichi[3]

1 Department of Geography, Geoinformatics and Meteorology, University of Pretoria, South Africa

2 South African Weather Service, Pretoria, South Africa

3 Department of Geomatics, Ahmadu Bello University, Zaria, Nigeria

*Address all correspondence to: u15221408@tuks.co.za

IntechOpen

References

[1] Patton AP, Laumbach R, Ohman-Strickland P, Black K, Alimokhtari S, Lioy PJ, et al. Scripted drives: A robust protocol for generating exposures to traffic-related air pollution. Atmospheric Environment. 2016;**143**: 290-299

[2] Gorai AK, Tchounwou PB, Mitra G. Spatial variation of ground-level ozone concentrations and its health impacts in an urban area in India. Aerosol and Air Quality Research. 2017;**17**(4):951-964

[3] Aliyu YA, Botai JO. An exposure appraisal of outdoor air pollution on the respiratory well-being of a developing city population. Journal of Epidemiology and Global Health. 2018; **8**(1):91-100

[4] Al-Awadi LT, Popov V, Khan AR. Seasonal effects of major primary pollutants in Ali Sabah Al-Salem residential area in Kuwait. International Journal of Environmental Technology and Management. 2015;**18**(1):54-82

[5] Marais EA, Jacob DJ, Wecht K, Lerot C, Zhang L, Yu K, et al. Anthropogenic emissions in Nigeria and implications for atmospheric ozone pollution: A view from space. Atmospheric Environment. 2014;**99**:32-40

[6] Aliyu YA, Musa IJ, Jeb DN. Geostatistics of pollutant gases along high traffic points in urban Zaria, Nigeria. International Journal of Geomatics and Geosciences. 2014;**5**(1): 19-31

[7] Aliyu YA, Botai JO. Reviewing the local and global implications of air pollution trend in Zaria, northern Nigeria. Urban Climate. 2018;**26**:51-59

[8] Snyder EG, Watkins TH, Solomon PA, Thoma ED, Williams RW, Hagler GSW, et al. The changing paradigm of air pollution monitoring. Environmental Science & Technology. 2013;**47**: 11369-11377

[9] Brook JR, Dann TF, Burnett RT. The relationship among TSP, PM_{10}, $PM_{2.5}$, and inorganic constituents of atmospheric participate matter at multiple Canadian locations. Journal of the Air & Waste Management Association. 1997;**47**(1):2-19

[10] Guo H, Wang Y, Zhang H. Characterization of criteria air pollutants in Beijing during 2014–2015. Environmental Research. 2017;**154**: 334-344

[11] Efe SI, Efe AT. Spatial distribution of particulate matter (PM_{10}) in Warri metropolis, Nigeria. The Environmentalist. 2008;**28**(4):385-394

[12] Yazdi MN, Delavarrafiee M, Arhami M. Evaluating near highway air pollutant levels and estimating emission factors: Case study of Tehran, Iran. Science of the Total Environment. 2015; **538**:375-384

[13] Aliyu YA, Botai JO. Appraising city-scale pollution monitoring capabilities of multi-satellite datasets using portable pollutant monitors. Atmospheric Environment. 2018;**179**:239-249

[14] Llanes S. How to calculate time-weighted average (TWA). In: 26th Annual California Industrial Hygiene Council (CIHC) Conference; 2016; San Diego, USA. Available from: http://www.thecohengroup.com/article/calculate-time-weighted-average-twa/ [Accessed: 07 October 2017]

[15] WHO. Evolution of WHO Air Quality Guidelines. Past, Present and Future. World Health Organization; 2017. Available from: http://www.euro.who.int/__data/assets/pdf_file/0019/331660/Evolution-air-quality.pdf [Accessed: 26 September 2017]

Chapter 5

Long-Distance LIDAR Mapping Schematic for Fast Monitoring of Bioaerosol Pollution over Large City Areas

Dimitar Stoyanov, Ivan Nedkov, Veneta Groudeva,
Zara Cherkezova-Zheleva, Ivan Grigorov, Georgy Kolarov,
Mihail Iliev, Ralitsa Ilieva, Daniela Paneva
and Chavdar Ghelev

Abstract

Light detection and ranging (LIDAR) atmospheric sensing is a major tool for remote monitoring of aerosol pollution and its propagation in the atmosphere. Combining LIDAR sensing with ground-based aerosol monitoring can form the basis of integrated air-quality characterization. When present, biological atmospheric contamination is transported by aerosol particles of different size known as bioaerosol, whose monitoring is now among the basic areas of atmospheric research, especially in densely-populated large urban regions, where many bioaerosol-emitting sources exist. Thus, promptly identifying the bioaerosol sources, including their geographical coordinates, intensities, space-time distributions, etc., becomes a major task of a city monitoring system. This chapter argues in favor of integrating a LIDAR mapping schematic with in situ sampling and characterization of the bioaerosol in the urban area. The measurements, data processing, and decision-making aimed at preventing further atmospheric contamination should be performed in a near-real-time mode, which imposes certain demands on the typical LIDAR schematics, including long-range sensing as a critical parameter, especially over large areas (10 – 100 km2). In this chapter, we describe experiments using a LIDAR schematic allowing near-real-time long-distance measurements of urban bioaerosol combined with its ground-based sampling and physicochemical and biological studies.

Keywords: LIDAR monitoring , particulate matter, atmospheric pollution, contaminations

1. Introduction

Atmospheric aerosol pollution or more appropriately particulate matter (PM) is key subject for the human health and ecosystem stability. At present, more than 2000 papers are published per year addressing research topics related to atmospheric aerosols [1]. However, surprisingly little is known regarding the genesis, composition, or dynamics of the atmosphere's microbial inhabitants, particle's chemical composition, particle's surface pollution, and their relation with the PM [2, 3]. In what concerns human health, the most harmful ones are the particles with sizes below 10 μm, standardized in terms of permissible concentration as $PM_{2.5}$ and PM_{10}. The fast pace of the information technology development in the first decade of twenty-first century created extraordinary possibilities for organizing systems for real-time monitoring of the atmosphere over urban areas. The implementation of such techniques requires the development of modern fast sensor systems, whose principle of operation, size, and convenient management ensures a problem-free functioning of the information systems without affecting the life in the particular urban area. The well-known techniques of PM monitoring include as a rule a small number of stations and the use of specialized algorithms for rapid assessment of the aerosol pollution's spatial structure above the city. These approaches suffer from a number of drawbacks, such as limited spatial resolution as compared with the urban structure, the low temporal resolution from the viewpoint of the quick polluting processes, and following the PM proliferation, the use of non-calibrated devices, and the subjectivity in selecting the sampling sites. A major drawback of the techniques in question is the relatively slow determination of the pollution sources. Of particular interest in assessing the air quality in urban areas is the biological contamination transported by PM in the form of bioaerosols. According to [4], almost 25% of the total airborne PM above land surfaces contains biological materials in the form of pollens, fungal spores, bacteria, viruses, and fragments from plants, animals, or living organisms with a size ranging from 0.02 up to 100 μm [5–8].

The present chapter discusses the capabilities of remote techniques of analyzing the pollution fields in conjunction with in situ sampling of bioaerosols and investigation of PM pollution as one of the promising approaches to raising the efficiency of systems of air monitoring over urban areas [9–11]. The light detection and ranging (LIDAR) techniques are considered as being among the ones best suited to real-time scanning and/or long-term monitoring of the pollution above large areas. Their advantages arise from several factors: (1) light wavelengths commensurate with or close to the PM size, (2) good resolution in terms of distance (5–30 m) and elevation angle (~1°), and (3) range of operation (20–30 km) [12]. In a scanning mode of operation, the LIDAR techniques allow one to construct LIDAR maps of the aerosol pollution distribution over large cities, which can be easily juxtaposed with the ground atmospheric monitoring networks. Special attention was paid here to the development of a specific LIDAR measurement schematic to be suitably combined with in situ bioaerosol sampling investigating its crystallo-chemical and surface structure, including the presence of bacteria and adsorbed nano- and micron-sized organic and inorganic contamination. The successful applicability of such combined methodology is demonstrated below by the results of a study of the PM size distribution and genesis in two typical urban zones within a large city (Sofia, capital of Bulgaria, population of approx. 1.3 million), in places where the LIDAR monitoring had established previously a mass concentration in peak hours of vehicle traffic.

1.1 LIDAR technologies applied in aerosol sensing

In the literature [13] one can find descriptions of various LIDAR techniques based on the interaction of laser radiation with bioaerosols, such as Mie, Rayleigh, and Raman [9, 10] scattering and laser-induced fluorescence of bioaerosols [14]. Depolarization LIDAR techniques are also employed, whereby measuring the laser light depolarization, and based on models of non-spherical aerosols, the latter's origin can be distinguished, e.g., dust, smoke, pollen, including bioagents as anthrax, ricin, etc.

The Mie scattering-based LIDAR techniques of bioaerosol remote sounding allow one to use simple and inexpensive LIDAR equipment efficient in long-distance vertical and horizontal scanning. Moreover, these LIDARs easily provide long-distance operation exceeding 30–50 km [9, 10] and 25 km by our LIDAR system [11]. A drawback of this approach arises when one needs a more detailed aerosol fields' characterization in terms of size distribution, non-sphericity or spectral properties of the particles. To satisfy such requirements, various more complex techniques had been developed. For example, UV laser-induced fluorescence allows one to assess the particles' elemental composition based on their electron and rotational-vibrational spectra [15]. The Raman IR spectroscopy technique provides information on the rotational-vibrational spectra of the molecular compounds constituting the particles. The differential absorption LIDAR (DIAL) techniques are also used to determine the presence of gaseous species in the atmosphere based on the absorption of specific laser light wavelengths [16].

The high degree of informativeness of these complex techniques from the viewpoint of a detailed bioaerosol description cannot, unfortunately, be combined with other requirements of importance for their efficient inclusion in networks for air-quality monitoring of the low atmosphere in urban areas. Here we will point out just the more important constraints: (i) high technical complexity, size, and cost of the LIDAR systems; (ii) short operative distances (2–5 km), seriously limiting the possibility of implementing bioaerosol monitoring system in large cities; and (iii) using the laser radiation focusing on the aerosol particles, etc.

The purpose of our investigations described below was to explore experimentally the possibility to make use of Mie scattering in a relatively simple LIDAR configuration emitting horizontally the near-ground atmosphere, thereby sounding from a single point the territory of a large city with high angular and distance resolutions. We present successively the LIDAR and sampling equipment in their concurrent functioning, as well as the results of the structural and biological studies, thus proving the capabilities of the LIDAR biomonitoring technique functioning within an urban air-quality monitoring and information system.

1.2 Methodology of combined LIDAR bioaerosol measurements

The methodology for performing the combined LIDAR bio-measurements, capable to be applied in near real-time regimes, is illustrated in **Figure 1**. In general its operation is based on the use of two independent subsystems: (i) LIDAR long-distance sensing subsystem, containing scanning LIDAR for fast 3D mapping aerosol fields (distance, time, mass concentration), and (ii) aerosol sampling equipment of the aerosol particles, transporting bio-contaminations disposed near the LIDAR illuminated resolution volume. The sampled bioaerosol particles are then processed by different methods and algorithms for determination of the mass concentration and LIDAR calibration, bio-contaminant characterization by different techniques, etc. The fast processing of calibrated LIDAR data provides opportunities for

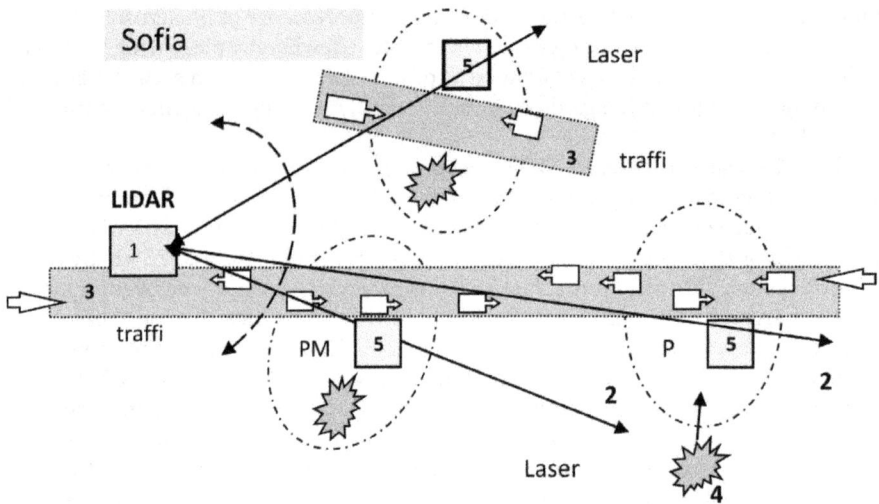

Figure 1.
Schematic diagram of a LIDAR bioaerosol measurement system.

well-timed decision-making for prevention of further spreading of bio-pollutants over the entire city area.

The LIDAR equipment is disposed in a single point (1). The laser beam is directed along specific paths (2) partly overlapping major city thoroughfares (3) with heavy traffic. These directions are selected on the basis of preliminary estimates (by, e.g., city monitoring network) of the presence of localized sources (4) emitting bio-contaminants that are subsequently transported in the near-ground atmosphere by PM. The PM sampling equipment (5) is placed close to the heavy-traffic spots, as explained below. In this schematic the scanning LIDAR capabilities could provide an effective coverage of too large areas of radius up to 25 km (1000 km^2 and more).

1.2.1 LIDAR sensing subsystem

LIDAR sensing subsystem for mapping the aerosol field above the sampling device (5) (**Figure 1**). An important point here is the contribution of the automobile traffic creating sufficiently strong backscattered LIDAR signals due to the PM, emitted by internal combustion engines. As these are formed at high temperatures, the possibility of its initially containing microorganisms is negligible. The hot PMs are then lifted up quickly in the air and enter the field of proliferation of bio-contaminants originating from the sources (4) (**Figure 1**). Thus, PM serves as a transport medium for the bioaerosols spreading over the city. In hours of heavy traffic, the backscattered LIDAR signal from bioaerosol-bearing PM is strong enough to allow remote sounding to distances of up to 10 km and more. After a fast computer processing, the LIDAR maps constructed can directly be used by the air-quality monitoring systems.

1.2.2 Sampling subsystem

It is described in detail below when we analyze the bio-contaminants found in the LIDAR sounding zones. The sampling is independent of the LIDAR subsystem, but it is important to synchronize the sampling time and duration with the LIDAR sounding in the vicinity of the sampling site. The LIDAR map can be used (with the aid of specialized algorithms) to establish the probable locations of sources of PM and bio- and other pollutants. The sampling data allow us to calibrate the LIDAR signal in terms of mass concentration (see below), thus shortening the time of reaction to the appearance of PM and other pollutants based on LIDAR observations only. We should note here that the delay due to processing the results as compared with the time necessary to process the LIDAR data (nearly real time) will not cause problems in the functioning of the entire system together with the city monitoring system. The sampling data will be saved and documented and could thus be used at a later stage.

2. Techniques and equipment

2.1 LIDAR subsystem

The LIDAR subsystem used in the experiments described here is of the scanning type with capability of scanning the entire hemisphere with an angular resolution of ~1° and is part of the EARLINET and ACTRIS LIDAR Station of Institute of Electronics—Bulgarian Academy of Sciences (IE-BAS) (see **Figure 2a**). It has been, and is, used in a large number of experiments of observation, monitoring, and mapping of the transport (including the transborder) of aerosol loadings over the city of Sofia [15]. The emitter is a CuBr vapor laser, with an output power of 2–4 W, pulse repetition rate of 5–10 kHz, and pulse duration of ~10 ns, the probing wavelength emitted being 510.6 nm. The receiving part operates in a photon-counting mode allowing sounding at distances up to R_{max} ~ 25–30 km [11]. The diameter of the Cassegrain-type receiving telescope is 19 cm. The LIDAR profiles are recorded and processed by a computer by the known algorithms [17, 18] and are then presented as multidimensional (depending on the sounding geometry) LIDAR maps. The laser beam divergence after collimation is within 1–2 mrad and can be varied if necessary. The receiving angle of the optical system can also be varied by varying the diameter of the diaphragm in front of the photon counter. Further, the spatial resolution is chosen within the 15–30 m range (usually 30 m); thus, the scattering volume is about twice as large as the volume pumped during the time of measurement by the aerosol sampler located nearby.

Figure 2b presents a vertical cross section of the atmosphere with origin at the LIDAR station and directed to downtown Sofia along the city's main thoroughfare, Tsarigradsko Chaussee Blvd. One clearly sees the vertical structure of the atmospheric aerosol density (in terms of backscatter coefficient), comprising a well-expressed ground layer with a height of 500–600 m, together with other

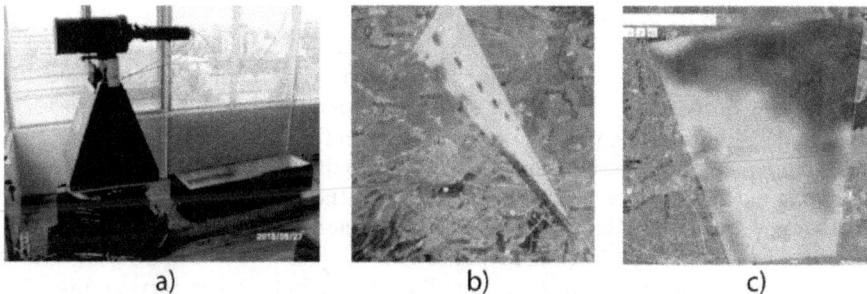

a)   b)   c)

Figure 2.
(a) Cu vapor LIDAR for atmospheric scanning; (b) LIDAR map of a vertical cross section of the atmosphere with origin at the IE-BAS LIDAR Station; (c) LIDAR map of the near-ground atmosphere of a city zone with high and low degree of PM pollution color-coded by brown and blue, respectively.

aerosol formations of limited size. The aerosol loading of the near-ground layer over the urbanized part of Sofia City Municipality is due to various sources (both local and transborder), with the main one being heavy automobile traffic, which is the subject of the present discussion. **Figure 2c** is a LIDAR map of part of the urban area, where one can see again well-defined zones of high (brown) and low (blue) MP pollution.

2.2 Calibration of LIDAR extinction and backscattering coefficients

Analyzing the PM aerosol loadings formed in the vicinity of heavy-traffic urban areas and experimentally measured by the LIDAR technique schematically presented in **Figure 2**, we were able to draw the important conclusion that the mass concentration of the aerosol loading of hot PM emitted by internal combustion engines is a key parameter when one applies a LIDAR-based methodology to air-quality monitoring in a large city. We thus calibrated the two major LIDAR parameters, namely, the extinction coefficient $\alpha(r)$ and the backscatter coefficient $\beta(r)$, in terms of the aerosol mass concentration following the well-known method [17, 19] and making use of the mass concentration M_a data obtained by the sampling device. For the LIDAR ratio $LiR = \alpha(r)/\beta(r)$, we adopted the typical value of $LiR = 50$ [16, 19]. The parameters $\beta(r)$ and $\alpha(r)$ were calculated using the LIDAR equation under the assumption of a horizontally homogeneous atmosphere:

$$P(r) = P_0 \frac{c\tau}{2} C \frac{\beta(r)}{r^2} exp\left[-2 \int_{r_0}^{r} \alpha(r)dr\right] \tag{1}$$

where $P(r)$ is the power of the detected laser radiation backscattered from the atmosphere from a distance $r = \frac{ct}{2}$ after a period of time t following the moment of laser pulse emission and τ is the pulse duration. Under the homogeneity assumption, the extinction coefficient $\alpha(r)$ is calculated as.

$$\alpha(r) = -\frac{1}{2}\frac{dS(r)}{dr}, \text{where } S(r) = \ln\left[r^2 P(r)\right] \tag{2}$$

Figure 3a and **b** presents the calibration dependencies of the mass concentration in [mg/m³] of, respectively, $\alpha(r)$ and $\beta(r)$. In both cases, the linear fit ($Y = A + B.x$) shows acceptable values of the standard deviation (less than 4%) and the correlation coefficient (over 0.92). The plots in **Figure 3** can be used directly for calibrating the LIDAR maps, shown above, in mass concentration.

2.3 LIDAR images of PM fields along the sounding directions

The city of Sofia is located in a valley surrounded by several mountains, which determine the meteorological conditions characterized by reduced possibility of self-cleaning of the atmosphere. Near-ground temperature inversions occur very often in the winter-spring transition period, on windless days, and in a stable atmosphere, with negative ecological effects due to the retention layer formed leading to increased concentration of pollutants in the boundary ground layer. The weather particularities motivated our choice of period of experimental observations, namely, February-March 2018 and 2019.

The object of our studies was the air pollution on windless days at two typical urban area points. (i) Intensive traffic (IT)—Tsarigradsko Chaussee Blvd. is the busiest thoroughfare in Sofia city with grade-separated dual carriageway in almost its entire length of 11.4 km, running from the northwest to the southeast.

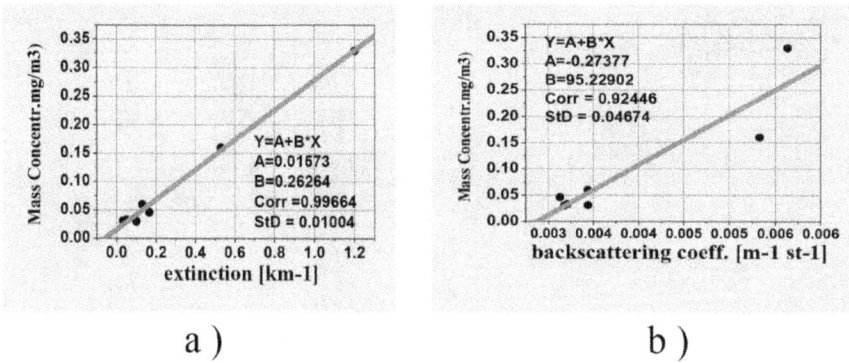

Figure 3.
Calibration plots for direct calculation of the aerosol mass concentration by both the extinction (a) and backscattering (b) coefficients.

The average intensity of the traffic according to Sofia Municipality is about 6000 vehicles per hour (Sofia Municipality Report 01.10.2017). The second sampling point was located at (ii) green area (GA) which was chosen at about 6.5 km (600 m ASL) from the LIDAR station on the roof of the Faculty of Biology of St. Kl. Ohridski University of Sofia; the building is located at a relatively busy thoroughfare, which forms a borderline between a green residential area and the largest city park.

Our LIDAR observation schedule complied with the generally accepted manner of measuring the aerosol mass concentration by air-quality monitoring systems. The sampling device pumps atmospheric air through the filter (typically a volume of 60–100 m^3) for an interval of about a few hours. Thus, the laser beam was stationary and directed to pass above the aspirator at a height of $h_{PM} < \delta R$, $\delta R \sim 30$ m being the LIDAR's radial resolution (point 1, **Figure 1**). The height of placing the aspirator was also chosen to comply with this condition, $h_{asp} < \delta R$. We thus could assume that we could neglect the vertical inhomogeneities of the atmospheric density. The LIDAR signals represent the number of backscattered photons $L_{phot}(k\delta R, \tau_m)$, where $k = 1..K_{max}$, $K_{max} = R_{max}/\delta R$, and $\tau_m = 5$ min are the time interval of photon accumulation. The total time of measurement lasted from 1 to several hours, depending on the particular weather situation. The computer system processes the input data by solving the LIDAR Eq. (1), with its output being profiles of the backscatter coefficient $\beta(k\delta R)$ or the extinction coefficient $\varepsilon(k\delta R)$, as calibrated in terms of aerosol mass concentration (see **Figure 3**). The set of LIDAR profiles obtained for the entire period of measurement is used to construct 3D LIDAR maps, with the x axis presenting the accumulation time with a step of $\tau_m = 5$ min and the y axis the distance from the LIDAR with a step δR. The z axis corresponds to the color-coded coefficients of backscatter or extinction. Thus, the position of the LIDAR station on the map has coordinates (0,0) at the start of measurements.

The series of figures below presents a set of such 3D LIDAR maps illustrating the aerosol loading space-time distribution within the region of LIDAR sounding and the aerosol sampler disposition. **Figure 4a** and **b** is a 3D LIDAR map of the backscatter coefficient distribution $\beta(k\delta R, j.\tau_m)$, $k = 1..K_{max}$, and $j = 1..J_{max}$; the vertical axis presents the distance to the LIDAR station with a step of $\delta R = 30$ m and $K_{max}\delta R > 10$ km, and the horizontal axis is the time (UTC) elapsed since the LIDAR sounding start (step of $\tau_m = 5$ min). **Figure 4a** and **b** is 3D LIDAR maps of the backscatter coefficient distribution $\beta(k\delta R, j.\tau_m)$, $k = 1..K_{max}$, and $j = 1..J_{max}$; the vertical axis presents the distance to the LIDAR station with a step of $\delta R = 30$ m and $K_{max}\delta R > 10$ km, and the horizontal axis is the time (UTC) elapsed since the LIDAR sounding starts (step of $\tau_m = 5$ min). The duration of sounding $\tau_{prob} = J_{max}.\tau_m$ exceeded 3 h.

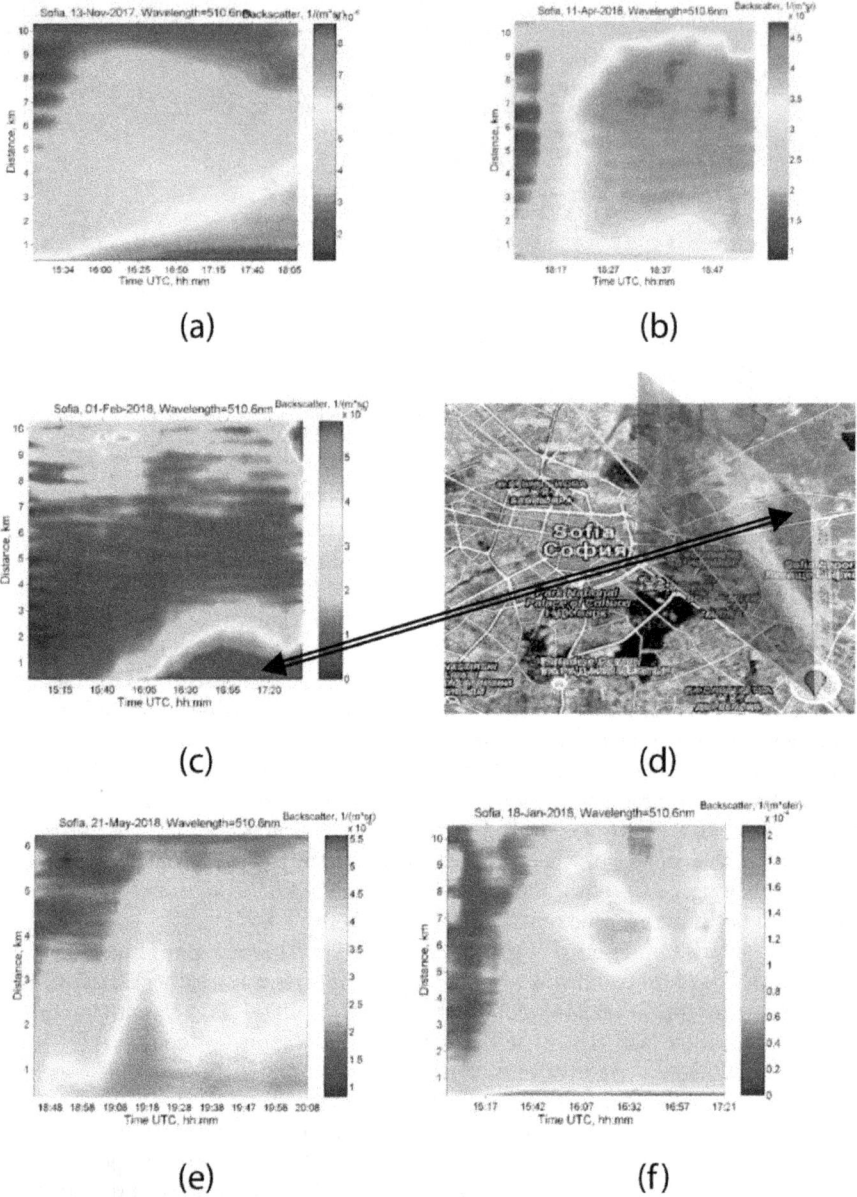

(a)

(b)

(c)

(d)

(e)

(f)

Figure 4.
(a and b) Hourly aerosol pollution loading (as 3D LIDAR maps) over the urban area with intensive traffic. LIDAR image (c) is overlapped in vertical position on the Google Maps (d) of the Sofia central part. Position of the LIDAR station is well seen. Such presentation is useful for simultaneous space-time tracing of the strong PM formation over the monitored area. (e) and (f) 3D LIDAR maps, demonstrating cases of complicated space-time dynamics of aerosol fields, transporting bioaerosol, emitted mainly by different local regions of the large city.

The backscatter coefficient $\beta(k\delta R, j\tau_m)$ is color-coded in terms of mass concentration, with the dark brown areas corresponding to the zones of the highest PM concentration and the light blue ones, to the lowest PM concentration. As seen, this approach allowed to follow the temporal behavior of the PM aerosol loading along the LIDAR sounding path to a distance of more than 10 km. For example, **Figure 4a** demonstrates that in the early afternoon hours, where the automobile traffic is relatively light, the aerosol loading is low, especially in the zones away from the downtown area. As the working day is nearing its end, the traffic intensity rises, and, correspondingly, the LIDAR response shows increasing mass concentration along the boulevard. The distances near the LIDAR station are characterized by very high mass concentration, since the LIDAR station (IE-BAS) is located close to the busy road junction—Tsarigradsko Chaussee Blvd. crossing with another four-lane boulevard. The image in **Figure 4b** illustrates the passing of a larger aerosol formation of length to the order of 8 km along the boulevard. In such cases we typically founded heavy bioaerosol loadings, emitted by the near disposed populated regions of limited areas of order of 1–2 km^2.

The images in **Figure 4c** illustrate a relatively small aerosol formation (as in **Figure 4b**), passing near the LIDAR station. The same image can be overlapped (in vertical disposition) on the Google Maps of the Sofia central part (**Figure 4c**) for better identification of the disposition of the emitted aerosol formation.

We should emphasize here the important point that the LIDAR maps can be used to follow the bioaerosol transport and to estimate the probable location of their sources. One could also draw the conclusion that, when the methodology discussed is employed, the sampling device location is not so critical in what concerns observing microorganisms in the air and establishing the contamination sources. The incorporation of the proposed LIDAR-based methodology into the city monitoring system could provide many additional advantages, such as forecasting the effects of selected bioaerosol loadings on the entire large city area, e.g., using modeling algorithms. The successful scanning by the LIDAR system in different directions (see **Figure 1**) could provide additional information for improving the local authorities' decision-making process.

2.4 Sampling and analytical methods

In the experiments presented, we directed the LIDAR beam to intense traffic area and green area to probe the near-surface atmosphere in a constant horizontal direction. The measurement time covered practically the entire period of the late afternoon traffic maximums, while maximums of the PM pollutions were clearly observed in the backscattered LIDAR signals, as received and processed by our system. The samples were taken in situ using a Hygitest 106 (Maimex), a high-efficiency portable device for sampling and concentration determination of PM in atmospheric aerosol. The apparatus allows simultaneously to take samples at four canals with a possibility to regulate the volume of air passing through the filters. The flow-rate of the aspirated air was measured by the sampling unit. The sampled volume was chosen to be smaller than the LIDAR resolution volume (typically ~250–300 m^3).

The dust was collected on a filter (borosilicon oxide) with #3 μm and #8 μm (FILTER-LAB, Material MCE, Lot.180509006 and 07). Also analytical filters #0.2 μm type FPP-15-2568-411-05795731-2008 were used, consisting of a layer of ultrathin threads (diameter of 1.5 μm) deposited on a piece of fabric and designed to collect aerosol particles of size exceeding 0.2 μm. Additionally, the material collected in situ on the filters after 3 h of aspiration during the LIDAR monitoring was studied by a number of methods: metal analysis (MA) has been carried out by devices and equipment—ICP-OES, PlasmaQuant PQ 9000 Elite (Analytik Jena), sample visualization, and PM morphological by scanning electron microscopy (SEM) and energy-dispersive and X-ray fluorescence systems (EDAX). Chemical composition, phase analysis, and particle size distribution were made using powder X-ray diffraction (XRD), photoelectron spectroscopy (XPS), and infrared (IR) spectroscopy. Phase identification was performed with the X-Pert program using ICDD-PDF2. Important tools for structural measurement was Mössbauer analysis which was made using apparatus Wissenschaftliche Elektronik GmbH, working with a constant acceleration mode: 57Co/Cr source, α-Fe standard. The sampling for investigation of microbiota in bioaerosols formed was achieved using a Hygitest 106 and axa replica technique on nutrient agar as well as Koch sedimentation method. Different elective media as nutrient agar; nutrient media for oligotrophs, actinomycetes, and fungi; blood agar; phenylethyl alcohol agar; MacConkey agar; and bacteria mobility-test medium [17] were used. Culturable bacteria isolated as pure cultures were identified by the methods of classical taxonomy [18] and the methods of the molecular taxonomy based on the PCR of 16S rDNA with universal eubacterial primers [20]. Molecular identification of the fungal strains was performed by PCR of the rDNA internal transcribed spacers (ITS) and primers ITS1 and ITS4 [21]. Positive PCR products were purified and sequenced (Macrogen Inc. Amsterdam, The Netherlands).

3. Particulate matter characterization

The natural aerosols are mixture of PM with highly varied crystallo-chemical structure bioaerosols, which are mixture from fungal spores, pollen, plant, etc. The particle research evaluation is a need for integrating the development of continuous monitoring technologies for determining both particle mass and chemical, physical, and biological methods for their identification. Such studies are of importance in determining the health and welfare effects of urban pollution and city transportation planning. As illustrated by **Figure 4**, we proved the possibility of a fast detailed remote analysis and monitoring of the air pollution over large urban regions, providing fast estimates of the air pollution transport over the city, as well as determination of pollution source location. Once a place of high PM concentration was localized by the LIDAR, samples were taken as described above. Initially, the particles were immobilized on single or multiple filters. This was followed by covering with a conductive carbon film deposited by sputtering of spectroscopically pure carbon in high vacuum. The experimental procedure allows for observing the particles' morphology and determining their mean size (**Figure 5a** and **b**) by means of SEM, EDAX, and MA.

The SEM images of the material collected on the filters after 3 h of aspiration during the LIDAR monitoring showed a large amount of particles larger than 2.5 μm and some amount of small particles (under 2.5 μm); most of the particles are included in quasi-aggregated structures, where nanosized particles could also be seen. **Figure 5a** illustrates the wide variety of quasi-spherical particles with an

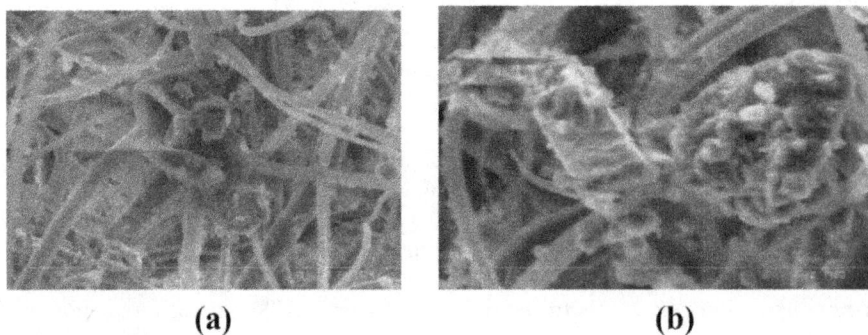

(a)   (b)

Figure 5.
SEM images of the particles under different magnification; the fibers reveal the filter structure of (a) PM with size ≥ 3 μm and (b) PM with size between 0.2 and 3 μm.

average size of about 2.5 μm. **Figure 5b** shows a typical shape of PM particles with sizes between 2.5 and 10 μm, which are agglomerates of hybrid origin.

3.1 Particulate matter distribution and metal's concentration analysis

Diameters of the PM varied from 20 nm to a more than 10 μm for pollen or plant debris [22–24]. After selecting typical particle images based on 10 points on the filter's surface, the percentage distribution of the particles with different sizes was obtained. The particles' size distribution study was conducted following the technique of random lines crossing particles of various diameters on optical microscopy and SEM images of immobilized PM particles under different magnification.

The SEM images of the material collected on the filters after 3 h of aspiration during the LIDAR monitoring showed a large amount of particles. The airborne PM can be divided into three classes (**Figure 6**), fine PM particles 2.5 μm in diameter or smaller, coarse PM particles 2.5–10 μm in diameter, and PM ≥ 10 μm, which differ not only in size but also in source, chemical composition, physical properties, and formation process. Major sources of $PM_{2.5}$ could be produced by motor vehicles, residential fireplace fossil fuel combustion by industry and wood stoves, vegetation burning, and smelting or other processes [25]. Our investigation for the period of 2 years showed that in the intensive traffic area, the majority of the particles are smaller than 10 μm as produced by automobile exhaust emissions, while in the green area, bigger particles appeared most probably from the wooded zone.

The studies conducted during the winter-spring transition period of 2018 and 2019 highlighted the alarming trend of increased content of the Cu, Fe, and Zn metals over the permissible concentration values [25]. **Figure 7** presents summarized data for two of the locations (blue is limited value, red dot is intensive traffic, and green dot is green area) that were objects of our studies considered here.

The above results were confirmed by EDAX measurements of PM collected by the functional filters. **Figures 8** and **9** demonstrate the presence of Cu and Fe. Less frequently, the presence of Pb was also noticed, again exceeding the permissible concentration values (**Figure 8**).

3.2 Crystallo-chemical structure of PM

Great attention in the physicochemical characterization of investigated PM samples was paid to compounds related to the observed overconcentrations of certain elements (**Figure 5**). The chemical elements are varying from main component level to trace elements. **Figure 10** shows the representative powder X-ray diffraction

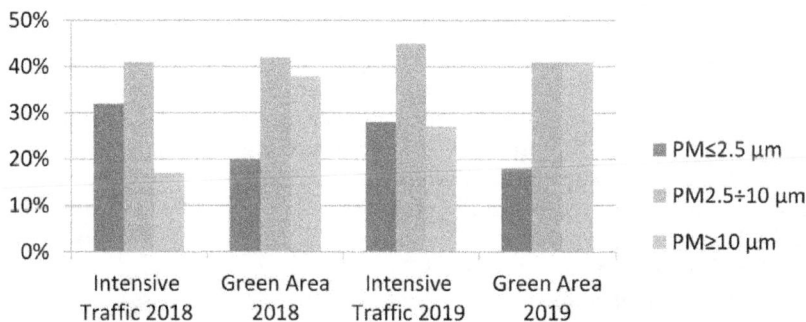

Figure 6.
Particle size distribution in two points in the urban area of Sofia City (for period of investigation February–April 2018 and 2019).

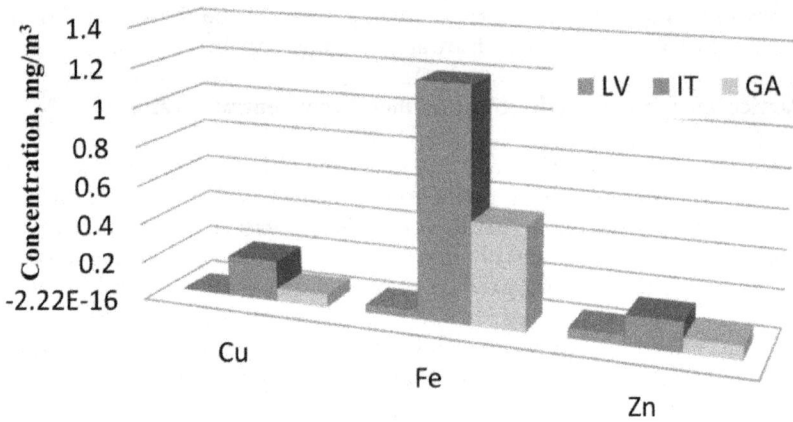

Figure 7.
Metals with concentrations exceeding the permissible limiting values (LV); the blue bars indicate the LV (annual average of mg/m³), and the red and green bars indicate the average concentrations in an intensive traffic (IT) and a green area (GA), respectively, as estimated by our LIDAR measurements.

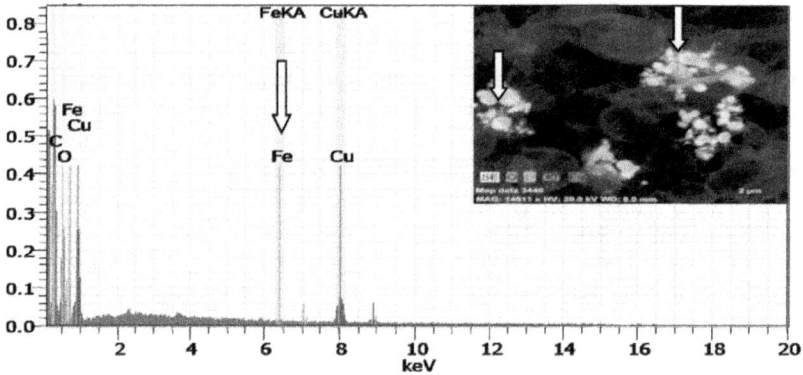

Figure 8.
EDS spectrum of element analysis and the insert presents SEM image of the element's distribution on the fixed particle adsorbed on the filters from IT zone (23.03.2019).

patterns of studied PM samples originated from different areas (IT and GA). The existence of low intensity and broad X-ray diffraction peaks laid on nonselective background in all registered patterns was observed. The main crystallite phases detected in X-ray diffractograms are silicates and aluminosilicates and carbonates. Lesser amounts of different sulfates were found also. Registered X-ray amorphous halos and nonselective background together with small intensity and high width of diffraction peaks of all registered crystallite phases indicate nearly amorphous structure, small particle size, and low crystallinity degree of PM material from both studied locations. The observations are in good consideration with obtained elemental composition and SEM analysis of studied materials. Regarding both analyses it could be concluded that PM greatly vary in size from nanometers to several tenths of micrometers. However, most of the particles on SEM images are aggregates of smaller particles (**Figure 8**).

Based on the obtained big iron content in studied PM, ^{57}Fe Mössbauer spectroscopy was applied to investigate samples. It allows to go deeper into the PM characteristics and to make more clear conclusions about the presented iron-bearing chemical compounds, their quantity, and dispersion. Represented

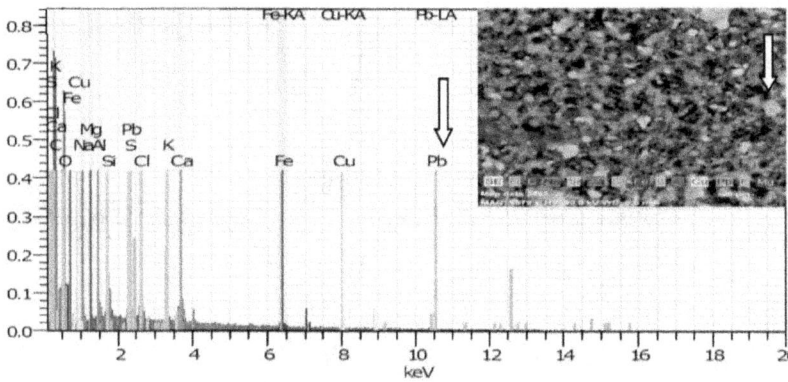

Figure 9.
EDS spectrum of element analysis and the insert presents SEM image of the element's distribution on the fixed particle adsorbed on the filters from IT zone (23.03.2019).

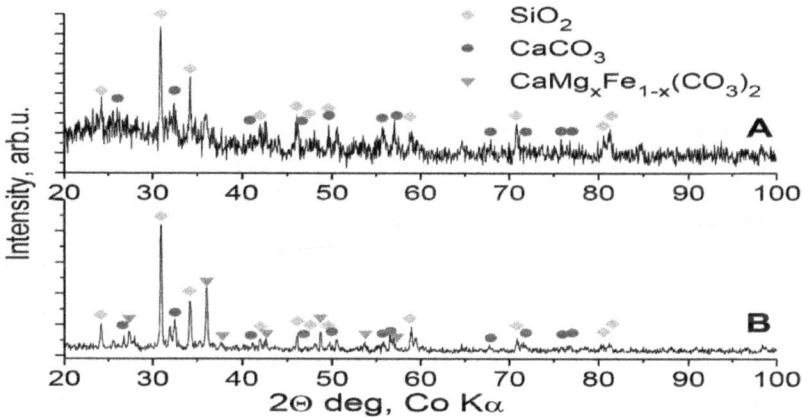

Figure 10.
Representative X-ray diffraction patterns of studied powder PM originated from IT (A) and GA (B).

Mössbauer spectra of samples can be seen on **Figure 11**. Sextet and doublet components were obtained after spectrum evaluation. They have the characteristic parameters of phases presented on the respective figures (**Figure 11A** and **B**). The main differences between studied PM from IT and GA could be regarded to quantity of presented phases and their particle size. After spectrum evaluation the sextet components with hyperfine parameters characteristic for spinel phase were resolved, highly non-stoichiometric magnetite $Fe_{3-x}O_4$ phase (or maghemite phase γ-Fe_2O_3) in all studied samples. Their quantity is much bigger in high-traffic area than in residence green area (29% vs. 12%). On the other hand, the calculated values of hyperfine effective fields of all registered magnetic phases (both spinel and hematite phase) are lower than the typical ones for the respective bulk phases. So it can be concluded that the particle size of these oxide phases is lower than 20 nm [26, 27]. Hematite phase dispersion is almost the same in IT and GA samples, but maghemite/magnetite phase particle size is smaller in IT area. It can be well seen on **Figure 11** that the doublet components are the main part of all PM spectra. According to the calculated hyperfine parameters of these doublets and comparison with previous investigations, it can be concluded that the majority of Fe-bearing compounds in IT PM are superparamagnetic (SPM) phases with nanometric size

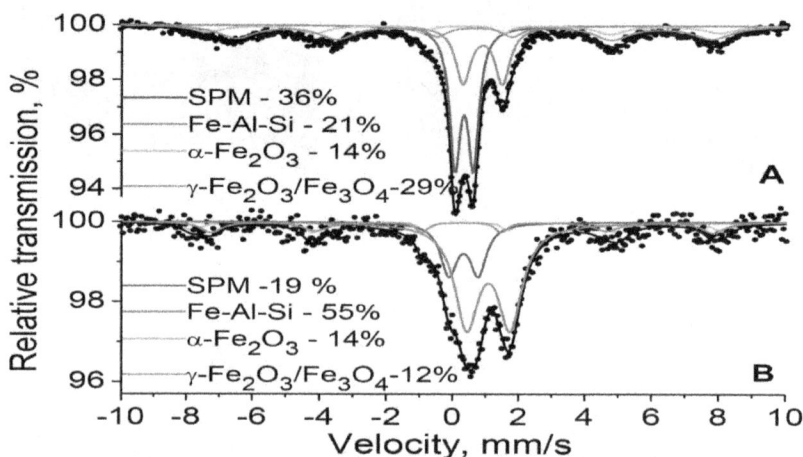

Figure 11.
Representative Mössbauer spectra of studied powder PM originated from IT (A) and GA (B)

(oxides or hydroxides). Fe^{3+} in paramagnetic phases as glass phases, sulfates, clay minerals, etc. is also presented in both locations (also in this component) [28]. Fe^{2+} was found in paramagnetic component indicative for the presence of aluminosilicate glass, ankerite, iron-containing carbonates or clay minerals, etc. [29]. The last component is the largest component in GA Möessbauer spectrum (see **Figure 11B**).

Comparative analysis of registered FTIR spectra of PM reveals appearance and increasing of bands typical for the organic compounds and inorganic salts (mainly sulfates and phosphates) typically presented on the surface of studied particles [30]. Obtained results indicate that silicates contributed to the highest percentage of total analyzed IR spectra signal for particles at both locations. Organic compounds are hydrocarbons and substituted hydrocarbons, adsorbed CO, etc. They could be regarded to partially oxidized products of fuel combustion. Registered X-ray photoelectron spectra of samples gave additional information for presented surface elements and their concentrations and chemical state. C, O, Si, Al, Ca, N, Na, and S have been detected on the PM surface. The determined binding energies are typical for Si-O, Al-Si-O, $Ca-CO_3$, C-H-N, C-H-NOx, C-H-N-Cl, Na-C-H-N-Cl, Cu-Cl-C-N-H-O, and Cu-O bonding, respectively [31]. Trace levels of Fe, Cu, Zn, and Cl were also found, and this could be related to automobile exhaust emissions and to brakes and tire wear. So, the analysis shows that the surface of investigated PM sample consists mainly of silicate and aluminosilicate compounds, as well as of different organic and inorganic carbon phases/carbonaceous species. XPS showed the presence of carbon black, which was attributed to the oxidative wear and subsequent deposition from related volatiles, as well as of graphitic particles emitted as a result of abrasive wear. The last one was registered in XRD patterns also. According to the literature data, most of the surface carbon comes from burning of fossil fuels, which is not surprising as all the samples were collected in winter period [32].

The analysis of obtained results using different characterization methods—XRD, Mössbauer, IR, and XPS—allows us to make conclusions about chemical composition of studied airborne PM, as well as about the quantity, crystallinity degree, and dispersion of compounds. The main phases registered in samples from both IT and GA are silicate, aluminosilicate, clay minerals, and sulfate compounds, as well as organic and inorganic (carbonate and coal) carbon phases. Elemental analysis showed that the Fe is the dominated metallic content in high-traffic area. The obtained bigger than usual content of iron in PM could be regarded to the airborne particles produced by transport and mainly by car engine performance [33]. This also explained the observation of higher iron content in IT samples compared to those from GA. On the other hand, Fe, Cu, Zn, Pb, and S in relatively high concentrations in all measured airborne PM (see **Figure** 7) could be regarded to abrasion of car brake lining material. Although the Pb is having been replaced in modern brake linings, the presence of cars older than 15–20 years is probably the explanation of registered Pb content in studied PM. This is an important issue for further regulations. Smaller quantity of other metals as Ba, Mg, Ni, and K has been found, which also could be considered as a result of car lining wear. Fe domination among trace elements was attributed to the easiest fragmentation of Cu, Zn, and Sn due to their lower mechanical properties, as well as their lower melting points, compared to steel and cast iron. Therefore, car braking could be considered to be one of the major sources of nonexhaust traffic-related emissions in all urban locations. The main source of silicate and aluminosilicate compounds could be considered to be mostly natural (silicate, aluminosilicate, clay minerals). The anthropogenic factors in PM formation are connected with street and house reparation activities. The noticeable concentrations of organic substances and elemental carbon have been recognized as a result of incomplete fuel combustion, lubricant volatilization during the combustion procedure, and residuals from the exhaust gases originating from power plants, small houses and different engines, road surface wear, etc. [34–37].

3.3 Bioaerosols

Bioaerosols may contain pollen, bacteria, actinomycetes, fungal spores, and sneezing and cough drops, as well as endotoxins, mycotoxins, and allergens. A number of studies have shown that bacteria in the air most commonly coexist with particulate matter and are thus transported over long distances [38, 39]. The average residence time of bioaerosols in the atmosphere may be from day to several weeks, depending on their size and aerodynamic properties [40]. Larger bioaerosol particles are retained in the upper airways of the human (oral and nasal cavities), while smaller ones can reach the lower pathways in the lungs [41, 42]. They can have a different negative effect on humans (infectious diseases, toxic effects, allergies, and even cancers). Most commonly, the symptoms and diseases resulting from the inhalation of bioaerosols are related to the respiratory system. Causes of some human infections, such as measles or tuberculosis, can spread through bioaerosols containing infectious microorganisms [43, 44]. Fungal spores, as part of bioaerosol particles, are most often associated with asthmatic symptoms and are a risk factor for various respiratory problems. Pulmonary plague caused by *Yersinia pestis* can spread after inhalation of bioaerosol particles containing the pathogen. Qualitative and quantitative composition of microorganisms varies greatly [2]. In the air over Erdemli, Turkey, during the passage of Saharan dust in March 2002, bacteria belonging to seven genera were isolated, and the majority of the species were referred to the genus *Streptomyces* [45]. In Bamako, Mali, representatives of 20 genera were identified during the passage of a large amount of desert dust, with *Bacillus* species representing 38% of all isolates, followed by genera *Kocuria* (12.8%), *Saccharococcus* (7.4%), and *Micrococcus* (6.4%). From the 95 species of bacteria identified in this study, about 10% are potential pathogens in animals, 5% are phytopathogens, and 25% are opportunistic human pathogens [46].

Eukaryotic microorganisms from genera *Cladosporium*, *Alternaria*, and *Epicoccum* are the dominant species found in open air in different parts of the world, while species of the genera *Penicillium* and *Aspergillus* are more often isolated from enclosed spaces [47]. Saharan sandstorms are responsible for the transmission of pathogens associated with widespread coral infections (predominantly *Aspergillus* genus) in the Caribbean region [48–50].

The concentration of bioaerosol particles varies greatly depending on the weather, location, and annual seasons. The wind, rain, solar radiation, and ozone are factors that influence the concentration of microorganisms in the air and may even have a bactericidal effect. The survival of bacteria in the air decreases with increasing temperatures as it begins to decrease when temperatures exceed 24°C [44–52]. High relative humidity (RH) can significantly reduce the bactericidal effect of ultraviolet light and hence increase the survival of bacteria [53]. Quantitative composition also influences some pollutants in the air, for example, formaldehyde, acrolein, ozone, and sulfur dioxide, which have a negative effect on the viability of the bacteria [54]. Due to the strong influence of these factors, the quantitative composition of the air microbiota is unstable and depends on local sources of pollution. Statistics based on the results of the US Environmental Protection Agency study show that bacterial concentration in open spaces is higher than in indoor pools [55]. The average concentration of all bacteria isolated from outdoor air is 10^2 CFU/m^3. Ninety-five percent of the culturable bacteria are mesophilic. In a study conducted in the United States, it has been found that the concentration of fungi is usually higher in open spaces than in the indoor. The highest concentration of fungi is measured in autumn and summer and the lowest in winter and spring. In open spaces, the concentration of fungi varies strongly from 1 to 8200 CFU/m^3 of air, with an average count of about 540 CFU/m^3 [56].

Fungi and their spores are more resistant to stress in the air environment than viruses and vegetative cells of the bacteria [44, 52]. Higher temperatures, wet substrates, and humidity provide favorable conditions for fungal development [57]. The quantitative analysis of the microbiome of air bioaerosols in the points tested reviled the presence of aerobic heterotrophic and oligotrophic microorganisms and fungi (**Tables 1** and **2**).

The analysis of the results obtained show that the quantities of the heterotrophic and oligotrophic microorganisms in the air of the first location mentioned are significantly higher. It is also obvious that the levels of these microorganisms in both points had a tendency to decrease during warm summer period, which correlated with lower urban traffic at this period. It must be noted that the quantity of fungi in the second location is significantly higher than the detected levels in the first one, but there is a tendency to increase during spring-summer period. The results are similar to those found by other authors [45, 46, 58–60]. Fifty-six pure cultures isolated from both points tested subjected to taxonomic investigation. More than 45% of the isolates are Gram-positive and belong to genus *Bacillus* (*B. cereus*, *B. pumilus*, *B. subtilis*, *B. megaterium*, *B. thuringiensis*, and *B. mycoides* are the dominant species). Information about the prevalence of such bacteria in similar sampling locations is available in the literature [39, 58, 60, 61].

Microorganisms	Quantity of the detected microorganisms CFU/m^3			
	Sampling January	Sampling May	Sampling August	Sampling November
Heterotrophic MO	64,256 ± 2.1	36,343 ± 2.4	32,147 ± 3.1	50,382 ± 1.5
Oligotrophic MO	56,950 ± 2.5	29,412 ± 1.8	27,876 ± 1.1	52,356 ± 2.3
Fungi	3747 ± 1.1	5223 ± 2.1	5322 ± 1.7	3245 ± 1.2

Table 1.
Quantitative analysis of culturable microorganisms in the air of the intense traffic.

Microorganisms	Quantity of detected microorganisms CFU/m³			
	Sampling January	Sampling May	Sampling August	Sampling November
Heterotrophic MO	9741 ± 1.1	6182 ± 2.1	5932 ± 1.8	83,462 ± 1.3
Oligotrophic MO	9086 ± 1.9	4777 ± 1.1	5323 ± 2.2	10,321 ± 2.1
Fungi	2248 ± 1.7	8992 ± 1.3	6332 ± 0.9	4672 ± 1.1

Table 2.
Quantitative analysis of culturable microorganisms in the air of the point green area.

The molecular analysis and sequencing confirmed these results. *Erwinia herbicola* was the dominant species from family *Enterobacteriaceae*. *Bacillus megaterium* and *Bacillus pumilus* as well as *Rathayibacter caricis*, *Arthrobacter* sp. *FXJ8.160*, *Acidovorax* sp. *NA2*, *Plantibacter flavus*, and *Kocuria rosea* were most frequently isolated. It can be summarized that from the group of Gram-positive bacteria in both locations, 48% of the isolates were related to *Bacillus* genus; 21% of all isolates were related to the *Enterobacteriaceae* family; 10% were related to the *Arthrobacter* genus; 4% were related to the genus *Exiguobacteria*; and 2% of the isolated microorganisms were related to the genera *Staphylococcus*, *Acidovorax*, *Plantibacter*, *Gordonia*, *Streptomyces*, *Kocuria*, and *Rathayibacter*. It is noteworthy that contingent pathogens (*Bacillus cereus*, *Bacillus pumilus*, *Erwinia herbicola*, *Enterobacter aerogenes*) are found among the isolates and such pathogens are also found in other studies of airborne microbial load [45]. Representatives of the genus *Gordonia* are conditionally pathogenic and isolated from sick patients. Representatives of the genus *Kocuria* are part of the resident microflora of the skin and mouth in humans but are also widespread in all elements of the environment. Representatives of the genus *Rathayibacter* cause serious damage to the nervous system.

Fungal isolates from both points investigated belong mainly to genera *Aspergillus* (*Aspergillus fumigatus*, *Aspergillus versicolor*), *Penicillium* (*P. sanguifluum*, *P. chrysogenum*, *P. brevicompactum*), *Cladosporium* (*C. sphaerospermum*), *Botrytis* (*B. cinerea*), and *Symmetrospora*. Some of these fungi isolated are typical human pathogens [47–49]. All these findings confirmed the idea that the investigations on particular matters (PM) must combine obligatory with analyzing of the microbiota in air aerosols.

4. Conclusions

The LIDAR monitoring methodology discussed opens up possibilities for rapid space-time identification and characterization of physicochemical and bio-pollution processes over the entire territory of large cities without affecting the normal living patterns of the city residents. Remote sensing allows one to locate the occurrence of pollution, which can be easily combined with in situ sampling. The experimental results described above clearly showed that at points of extreme pollution, the small-size particles (less than 2.5 μm) predominated in areas of heavy traffic, while nanosized PM were also found. Further, in intensive traffic areas, we observed metal particles, whose concentrations exceeded the maximal admissible levels. LIDAR sounding of "greener" areas revealed the predominant presence of particles with sizes ≥10 μm; the analyses conducted indicated their prevailing biological origin. The LIDAR maps created can be further used for tracing the full air-mass transport, carrying contamination from a number of pollution sources (chemical, biological, dust, etc.) distributed over the scanned region. Finally, we should emphasize the simplicity of the LIDAR and the aerosol sampling equipment used and, thus, the possibilities for its wide use in any populated region, where keeping the air quality within tolerable levels is problematic. We should once again note that methodology developed affects negligibly the residents' lifestyle in urban regions. Therefore, it is our belief that it offers promising paths for wide application in air-quality monitoring in highly polluted large urban areas.

Acknowledgements

This work was financed in part by contract DH18/16 with the National Science Fund, Bulgaria, and included in the European Program of the COST Action CA16202. The scanning LIDAR system was developed as part of the EARLINET and ACTRIS-2, Horizon 2020 EU projects.

Author details

Dimitar Stoyanov[1], Ivan Nedkov[1*], Veneta Groudeva[2], Zara Cherkezova-Zheleva[3], Ivan Grigorov[1], Georgy Kolarov[1], Mihail Iliev[2], Ralitsa Ilieva[2], Daniela Paneva[2] and Chavdar Ghelev[1]

1 Institute of Electronics, Bulgarian Academy of Sciences, Sofia, Bulgaria

2 Faculty of Biology, St. Kliment Ohridski University of Sofia, Sofia, Bulgaria

3 Institute of Catalysis, Bulgarian Academy of Sciences, Sofia, Bulgaria

*Address all correspondence to: nedkovivan@yahoo.co.uk

IntechOpen

References

[1] Fuzzi S, Baltensperger U, Carslaw K, Decesari S, Denier van der Gon H, Facchini MC, et al. Particulate matter, air quality and climate: Lessons learned and future needs. Atmospheric Chemistry and Physics. 2015;**15**:8217-8299. DOI: 10.5194/acp-15-8217-2015

[2] Brodie EL, De-Santis TZ, Moberg Parker JP, Zubietta IX, Piceno YM, Andersen GL. Urban aerosols harbor diverse and dynamic bacterial populations. Proceedings of the National Academy of Sciences of the United States of America. 2007;**104**(1):299-304. DOI: 10.1073/pnas.0608255104

[3] Jones AM, Harrison RM. The effects of meteorological factors on atmospheric bioaerosol concentrations—A review. Science of the Total Environment. 2004;**326**(1-3):151-180. DOI: 10.1016/j.scitotenv.2003.11.021

[4] Macher J. Bioaerosols: Assessment and Control, American Conference of Governmental Industrial Hygienists. USA: OH Cincinnati; 1999. 322 p. ISBN: 978-1-882417-29-1

[5] Stanley RG, Linskins HF. Pollen: Biology, Chemistry and Management. 1st ed. Berlin, Germany: Springer Verlag; 1974. DOI: https://www.springer.com/gp/book/9783642659072

[6] Gregory PH. The Microbiology of the Atmosphere. 2nd ed. London: Hall; 1973. 377 p. DOI: https://catalogue.nla.gov.au/Record/395639

[7] Maricovich H, editor. Black's Medical Dictionary. 42nd ed. USA: A&C Black; 2009. p. 765. ISBN-10:9780713689020

[8] Measures RM. Laser Remote Sensing: Fundamentals and Applications. 1st ed. NY, USA: Wiley&Sons; 1984. 510 p. DOI: https://www.worldcat.

org/title/laser-remote-sensing-fundamentals-and-applications/oclc/123159913

[9] Kovalev VA, Eichinger WE. Elastic LIDAR: Theory, Practice, and Analysis Methods. 1st ed. NY, USA: Wiley&Sons; 2004. 615 p. DOI: 10.1002/0471643173

[10] Weitkamp C, editor. LIDAR Range-Resolved Optical Remote Sensing of the Atmosphere. Springer Series in Optical Sciences, Springer; 2005. 456 p. DOI: 10.1007/b106786

[11] Stoyanov D, Dreischuh T, Grigorov I, Kolarov G, Deleva A, Peshev Z, et al. Near surface aerosol LIDAR mapping of Sofia Area. On the synergy with city sensor network. In: Proceedings of the Final Meeting—Sixth Sci. Meeting EuNetAir. 2016. pp. 61-64. DOI: 10.5162/6 EuNetAir2016/16

[12] Simard JR, Roy G, Mathieu P, Larochelle V, McFee J, Ho J. Standoff Integrated bioaerosol Active Hyperspectral Detection (SINBAHD): Final Report. 2002: DREV-TR-2002-125. DOI: pubs.drdc-rddc.gc.ca/BASIS/pcandid/www/engpub/DDW?W%3DSYSNUM=518849

[13] Simard JR, Roy G, Mathieu P, Larochelle V, McFee J, Ho J. Standoff sensing of bioaerosols using intensified range-gated spectral analysis of laser induced fluorescence. IEEE Transactions on Geoscience and Remote Sensing. 2004;**42**(4):865-874. DOI: 10.1109/TGRS.2003.823285

[14] He TY, Stanič S, Gao F, Bergant K, Veberič D, Song Q, et al. Tracking of urban aerosols using combined LIDAR-based remote sensing and ground-based measurements. Atmospheric Measurement Techniques. 2012;**5**: 891-900. DOI: 10.5194/amt-5-891-2012

[15] Peshev ZY, Dreischuh TN, Toncheva EN, Stoyanov DV. Two-wavelength

LIDAR characterization of atmospheric aerosol fields at low altitudes over heterogeneous terrain. Journal of Applied Remote Sensing. 2012;**6**(1):063581. DOI: 10.1117/1.JRS.6.063581

[16] Klett J. Stable analytical inversion solution for processing LIDAR returns. Applied Optics. 1981;**20**(1):211-220. DOI: 10.1364/AO.20.000211

[17] Atlas KM. Handbook of Microbiological Media. 4th ed. Washington DC: ASM Press and Roca Raton, London, New York, USA: CRC Press; 2010. 2040 p. ISBN-10: 9781439804063

[18] Whitman W, editor. Bergey's Manual of Systematics of Archaea and Bacteria (BMSAB). Bergey's Manual Trust Publ; 2015. 2011 p. ISBN:9781118960608. DOI: 10.1002/978111896060

[19] Fernald F. Analysis of atmospheric LIDAR observations: Some comments. Applied Optics. 1984;**23**(5):652-653. DOI: 10.1364/AO.23.000652

[20] Wilson KH, Blitchington RB, Greene RC. Amplification of bacterial 16S ribosomal DNA with polymerase chain reaction. Journal of Clinical Microbiology. 1990;**28**(9):1942-1946. DOI: jcm.asm.org/content/28/9/1942

[21] White TJ, Bruns T, Lee S, Taylor JW. Amplification and direct sequencing of fungal ribosomal RNA genes for phylogenetics. In: Innis MA, Gelfand DH, Sninsky JJ, White TJ, editors. PCR Protocols: A Guide to Methods and Applications. New York.: Academic Press, Inc.; 1990. pp. 315-322. DOI: 10.1016/0307-4412(91)90165-5

[22] Fröhlich-Nowoisky DAP, Després VR, Pöschl U. High diversity of fungi in air particulate matter. Proceedings of the National Academy of Sciences of the United States of America.

2009;**106**(1):12814-12819. DOI: 10.1073/pnas.0811003106

[23] Jaenicke R. Abundance of cellular material and proteins in the atmosphere. Science. 2005;**308**(5718):73. DOI: 10.1126/science.1106335

[24] Hind WC. Aerosol Technology: Properties, Behavior, and Measurement of Airborne Particles. 2nd ed. NY: John Wiley&Sons; 2012. 504 p. ISBN: 978-1-118-59197-0

[25] World Health Organization (WHO). Health Effects of Particulate Matter. Copenhagen: Regional office for Europe; 2013. ISBN: 978 92 890 0001 7

[26] Fock J, Hansen MF, Frandsen C, Mørup S. On the interpretation of Mössbauer spectra of magnetic nanoparticles. Journal of Magnetism and Magnetic Materials. 2018;**445**:11-21

[27] De Grave E, Van Alboom A. Evaluation of ferrous and ferric Mössbauer fractions. Physics and Chemistry of Minerals. 1991;**18**:337-342. DOI: 10.1007/BF00200191

[28] Mahieu B, Ladrière J, Desaedeleer G. Mössbauer spectroscopy of airborne particulate matter. Journal de Physique Colloques. 1976;**37**(C6):C6-837-C6-840. DOI: 10.1051/jphyscol:19766176

[29] Gietl J, Lawrence R, Thorpe A, Harrison R. Identification of brake wear particles and derivation of a quantitative tracer for brake dust at a major road. Atmospheric Environment. 2010;**44**:141-146. DOI: 10.1016/j.atmosenv.10.016

[30] Ji Z, Dai R, Zhang Z. Characterization of fine particulate matter in ambient air by combining TEM and multiple spectroscopic techniques—NMR, FTIR and Raman spectroscopy. Environmental Science: Processes & Impacts. 2015;**17**:552-560. DOI: 10.1039/C4EM00678J

[31] Moulder F, Sticke WF, Sobol PE, Bombel KD, Castain J, editors. Handbook of X-ray Photoelectron Spectroscopy. 2nd ed. Waltham, USA: Perkin-Elmer Corporation, Physical Electron Division; 1992. DOI: 10.1002/sia.740030412

[32] González L, Longoria-Rodríguez F, Sánchez-Domínguez M, Leyva-Porras C, Acuña-Askar K, Kharissov B, et al. Seasonal variation and chemical composition of particulate matter: A study by XPS, ICP-AES and sequential microanalysis using Raman with SEM/EDS. Journal of Environmental Sciences. 2018;**74**:32-49. DOI: 10.1016/j.jes.2018.02.002

[33] Thorpe A, Harrison RM. Sources and properties of non-exhaust particulate matter from road traffic: A review. Science of the Total Environment. 2008;**400**:270-282. DOI: 10.1016/j.scitotenv.2008.06.007

[34] Kelly F, Fussell J. Size, source and chemical composition as determinants of toxicity attributable to ambient particulate matter. Atmospheric Environment. 2012;**60**:504. DOI: 10.1016/j.atmosenv.2012.06.039

[35] Chen H, Laskin A, Baltrusaitis J, Gorski CA, Scherer MM, Grassian VH. Coal fly ash as a source of iron in atmospheric dust. Environmental Science & Technology. 2012;**46**(4):2112-2120. DOI: https://www.dora.lib4ri.ch/eawag/islandora/object/eawag:7063

[36] Kukutschová J, Moravec P, Tomásek V, Matejka V, Smolík J, Schwarz J, et al. On airborne nano/micro-sized wear particles released from low-metallic automotive brakes. Environmental Pollution. 2011;**159**:998-1006. DOI: 10.1016/j.envpol.2010.11.036

[37] Cherkezova-Zheleva Z, Paneva D, Kunev B, Kolev H, Shopska M, Nedkov I. Challenges at characterization of particulate matter—A case study. Bulgarian Chemical Communications. 2018;**50F**:93-98. DOI: http://www.bcc.bas.bg/index.html

[38] Schlesinger P, Mamane Y, Grishkan I. Transport of microorganisms to Israel during Saharan dust events. Aerobiologia. 2006;**22**:259-273. DOI: 10.1007/s10453-006-9038-7

[39] Maki T, Puspitasari F, Hara K, Yamada M, Kobayashi F, Hasegawa H, et al. Variations in the structure of airborne bacterial communities in a downwind area during an Asian dust (Kosa) event. Science of the Total Environment. 2014;**488-489**:75-84. DOI: 10.1016/j.scitotenv.2014.04.044

[40] De Nuntiis P, Maggi O, Mandrioli P, Ranalli G, Sorlini C. Monitoring the biological aerosol. In: Madrioli P, Caneva G, Sabbioni C, editors. Cultural Heritage and Aerobiology. Dordrecht: Kluwer Academic Publishers; 2003. pp. 107-144. ISBN: 978-94-017-0185-3

[41] Davies A, Thomson G, Walker J, Bennett A. A review of the risks and disease transmission associated with aerosol generating medical procedures. Journal of Infection Prevention. 2009;**10**(4):122-126. DOI: 10.1177/1757177409106456

[42] Després VR, Huffman JA, Burrows SM, Hoose C, Safatov AS, Buryak G, et al. Primary biological aerosol particles in the atmosphere: A review. Tellus B: Chemical and Physical Meteorology. 2012;**64**(1):2-20. DOI: 10.3402/tellusb.v64i0.15598

[43] Jones RM, Brosseau LM. Aerosol transmission of infectious disease. Journal of Occupational and Environmental Medicine. 2015;**57**(5):501-508. DOI: 10.1097/JOM.0000000000000448

[44] Ijaz MK, Zargar B, Wright KE, Rubino JR, Sattar SA. Generic aspects of the airborne spread of human

pathogens indoors and emerging air decontamination technologies. American Journal of Infection Control. 2016;**44**(9 Suppl):S109-S120. DOI: 10.1016/j.ajic.2016.06.008

[45] Griffin DW, Kubilay N, Kocak M, Gray MA, Borden TC, Kellogg CA, et al. Airborne desert dust and aeromicrobiology over the Turkish Mediterranean coastline. Atmospheric Environment. 2007;**41**:4050-4062. DOI: 10.1016/j.atmosenv.2007.01.023

[46] Kellogg CA, Griffin DW, Garrison VH, Peak KK, Royall N, Smith RR, et al. Characterization of aerosolized bacteria and fungi from desert dust events in Mali, West Africa. Aerobiologia. 2004;**20**:99-110. DOI: 10.1023/B:AERO.0000032947.88335.bb

[47] Akerman M, Valentine-Maher S, Rao M, Taningco G, Khan P, Tuysugoglu G, et al. Allergen sensitivity and asthma severity at an inner city asthma center. The Journal of Asthma. 2003;**40**(1):55. DOI: 10.1081/JAS-120017207

[48] Shinn EA, Smith GW, Prospero JM, Betzer P, Hayes ML, Garrison V, et al. African dust and the demise of Caribbean coral reefs. Geophysical Research Letters. 2000;**27**(19):3029-3032. DOI: 10.1029/2000GL011599

[49] Weir-Brush JR, Garrison VH, Smith GW, Shinn EA. The relationship between gorgonian coral (Cnidaria: Gorganacea) diseases and African dust storms. Aerobiologia. 2004;**20**(2):119-126. DOI: 10.1023/B:AERO.0000032949. 14023.3a

[50] Wilken JA, Sondermeyer G, Shusterman D, McNary J, Vugia DJ, McDowell A, et al. Coccidioidomycosis among workers constructing solar power farms, California, USA, 2011-2014. Emerging Infectious Diseases. 2015;**21**(11):1997-2005. DOI: 10.3201/eid2111.150129

[51] Haig CW, Mackay WG, Walker JT, Williams C. Bioaerosol sampling mechanisms, bioefficiency and field studies. The Journal of Hospital Infection. 2016;**93**(3):242-255. DOI: 10.1016/j.jhin.2016.03.017

[52] Tang JW. The effect of environmental parameters on the survival of airborne infectious agents. Journal of the Royal Society Interfac. 2009;**6**(Suppl):S737-S746. DOI: DOI. 10.1098/rsif.2009.0227.focus

[53] Peccia J, Werth HM, Miller S, Hernandez. Effects of relative humidity on the ultraviolet induced inactivation of airborne bacteria. Aerosol Science and Technology. 2001;**35**(3):728-740. DOI: 10.1080/02786820152546770

[54] Won E, Ross H. Reaction of Airborne Rhizobium meliloti to some environmental factors. Applied Microbiology. 1969;**18**:556-557

[55] Tsai FC, Macher JM. Concentrations of airborne culturable bacteria in 100 large US office buildings from the BASE study. Indoor Air. 2005;**15**(Suppl 9):71-81. DOI: 10.1111/j.1600-0668.2005.00346

[56] Viegas C, Viegas S, Gomes A, Täubel M, Sabino R, editors. Exposure to Microbiological Agents in Indoor and Occupational Environments. 1st ed. Springer International Publishing AG; 2017. 415 p. DOI: 10.1007/978-3-319-61688-9

[57] Tang W, Kuehn TH, Simcik MF. Effects of temperature, humidity and air flow on fungal growth rate on loaded ventilation filters. Journal of Occupational and Environmental Hygiene. 2015;**12**(8):525-537. DOI: 10.1080/15459624.2015.1019076

[58] zur Nieden HA, Jankofsky M, Stilianakis NI, Boedeker R-H, Eikmann TF. Effects of bioaerosol polluted outdoor air on airways of residents:

A cross sectional study. Occupational and Environmental Medicine. 2003;**60**(5):336-342. DOI: 10.1136/oem.60.5.336

[59] Kellogg CA, Griffin DW. Aerobiology and the global transport of desert dust. Trends in Ecology & Evolution. 2006;**21**(11):638-644. DOI: 10.1016/j.tree.2006.07.004

[60] Bragoszewska E, Mainka A, Pastuszka JS. Concentration and size distribution of culturable bacteria in ambient air during spring and winter in Gliwice: A typical urban area. Atmosphere. 2017;**8**(12):239-252. DOI: 10.3390/atmos8120239

[61] Shivaji S, Chaturvedi P, Suresh K, Reddy GS, Dutt CB, Wainwright M, et al. *Bacillus aerius* sp. nov., *Bacillus aerophilus* sp. nov., *Bacillus stratosphericus* sp. nov. and *Bacillus altitudinis* sp. nov., isolated from cryogenic tubes used for collecting air samples from high altitudes. International Journal of Systematic and Evolutionary Microbiology. 2006;**56**(Pt 7):1465-1473. DOI: 10.1099/ijs.0.64029-0

Chapter 6

Smart Environment Monitoring System Using Wired and Wireless Network: A Comparative Study

Tabbsum Mujawar and Lalasaheb Deshmukh

Abstract

This chapter focuses on the implementation of a smart environment monitoring system using wired and wireless sensor networks (WSN). The goal was to develop a LabVIEW based system to monitor environmental parameters that provide inaccessible, real-time monitoring. The development of portable and efficient environment monitoring system based on LabVIEW GUI that monitors various environmental parameters such as temperature, relative humidity, Air quality and light intensity was developed. This chapter targets on both wired and wireless approach for environment monitoring. The limitations of wired network were explained by flourishing the portable system. For proceedings with the impediment and insufficiency of wired network, Arduino augmentation ascendancy, are mingled with XBee wireless sensor network. The data from the environment was sent to the sink node wirelessly through mote. Monitoring of the data was done in a personal computer (PC) through a graphical user interface made by LabVIEW. The pertinent sensor for each was connected to analog input of Arduino UNO and their values are displayed on front panel of LabVIEW. LabVIEW run time engine makes the system cost effective and facile. To reveal the effectiveness of the system, some measurement results are also predicted in this chapter.

Keywords: wired network, wireless sensor network, LabVIEW, web publishing tool

1. Introduction

Environment monitoring plays an important role in all the sectors. It is a forthcoming relevance field which is of fastidious rate to our country. Metropolitan cities with superior absorption of industry, rigorous transportation and soaring population mass are major sources of air pollution, which results in monitoring of environment. To think about the environment, it has turned out to be one of the prime concerns for almost every country in the world. Due to enormous increased in industrialization, the recent condition is obviously altering towards more environment gracious solutions.

This chapter discusses the different environmental and air quality parameters using respective sensors for it and provides various opportune services for users who can administer the information via a website from long-distance. This system comprises of both wired and wireless networks. Wired communications is a wide name used to portray the communication process that utilizes the cables and wiring to convey the data. Usually, wired communications are appreciated widely by research community due to its stability in services. They are not influenced by external environmental effects as compared to wireless networks. For some services, the potency and pace of the communication is finer to other solutions, such as satellite. Due to this distinctiveness, wired connections linger trendy, yet wireless system sustained to proceed. Environment monitoring with wired network have some limitations such that wired sensors could not be implemented in remote areas. Also it is very complex and costly to mount and sustain the wired networks [1]. Additionally, if a wire between the two devices gets breaks, the communication between these two gets collapsed; hence, the entire network will also fail.

Letter the initiative of replacing the wired with wireless network was brought and it overcomes approximately all the troubles with the wired communication nevertheless it hold disadvantages of the sluggish bandwidth and growth of interference. Wireless communications is a rapidly increasing technology endow with the litheness and mobility in our environment. The noticeable benefit of wireless transmission is a key diminution and simplification in wiring. The cabling cost in industrial installations is 130–650 US$ per meter and using wireless technology, it would be eradicated around 20–80% [1]. The skillful organization of the equipment through efficient monitoring of the environment augments an additional hoard in terms of cost, e.g., Wang et al. [2]. The wireless system developed by Honeywell to scrutinize steam traps saves the total cost effectively about 100,000–300,000 US$ annually [2]. The impracticable sensor applications' technology, viz. monitoring far-off areas and locations, is featured with unrestricted mechanism and litheness for sensors and augmented the network heftiness. Moreover, WSN technology makes the system reliable and less costly. It allows more rapidly exploitation and deployment of different sensors because this network provides various properties to the sensor nodes. Further, an integration of WSN technology with MEMS makes the motes with enormously stumpy cost, miniature sized and least power. MEMS are the inertial sensors, pressure sensors, temperature sensors, humidity sensors, strain-gage sensors and various piezo and capacitive sensors for proximity. Over the last decade, the technology of wireless sensor network (WSN) has been widely used in many real time applications and these miniaturized sensors can sense, process and communicate. Most wireless sensor nodes are capable of measuring temperature, acceleration, light, illumination, humidity; level of gases and chemical materials in the surrounding environment.

WSN is a compilation of wireless sensor nodes. A WSN is also an amalgamation of an integer of motes with limited communication ability. The co-ordination between the sensor nodes provides ability to process and to gather information in a large amount [3, 4]. Also, ad-hoc networks can be created. Generally, WSN networks are categorized in two types: structured and unstructured. In unstructured WSN, the sensor nodes are deployed in an ad-hoc manner without any careful planning. Once nodes are deployed, monitoring and processing of data is done in unattended environment. In structured WSN, motes are deployed in preplanned approach. The structured wireless sensor network is superior to unstructured one, because cost and maintenance required to deploy the node are less. The nodes in structured WSN are positioned at exact locations to offer coverage, whereas unstructured deployment has uncovered areas. Wireless sensor network aims to give co-ordination among the physical conditions and the internet globe. It has the following features:

- WSN should be reliable

- More accurate

- Flexible in nature

- Cost effective

- Easy to install.

Tilak et al. [5] have shown that the intellectual sensors can gather data from disaster area, floods and also from revolutionary attacks. The network is promising for

- Collection of information

- Dealing out of information easily and

- Environment monitoring for numerous applications.

Due to the above compensation WSN becomes vital part of near future applications.
By using WSN based environmental monitoring system it is possible to transform the customarily environmental monitoring methods. Conventionally data loggers were used to collect the data from environment and this was time overriding and fairly costly. To avoid the drawbacks of this system, we developed a system which is portable and cost effective. The LabVIEW (Laboratory Virtual Instrument Engineering Workbench) and an Arduino IDE are the programming tools used for this system. But, the writing programming is mostly used in Arduino [6]. Meanwhile, LabVIEW uses a programming type of block diagram. In the present system, it is decided to use the Arduino platform or microcontroller for the deployment of WSN nodes. This is an embedded board having included USB competence. The miniature and user responsive nature makes it more superior than other advanced microcontrollers. These microcontrollers have more on chip facilities such as +5 V, analog and digital pins. It does not have on board power jack. Due to the auto switching capability of ATMega 328 microcontroller, no external power jumpers are required. The use of an Arduino simplifies the process of working with microcontrollers and additionally it offers some advantages to the users over other systems such as cross-platform, simple, clear programming environment, open source and extensible software and hardware. Arduino platform has good specifications e.g. cheap, easy to use and wide varieties of shields that have been emerged with many different purposes such as Ethernet and GSM support. While, if we want to create a multifunction code for carrying out on multicore processors this would be possible using LabVIEW tool. It has a graphical palette to create and run VI's. Any complex programming can be done easily using the tools available in this software. This environment monitoring system uses the web publishing tool to display the monitored data on the web page for remote monitoring. In this chapter, we studied both wired and wireless environment monitoring system.

2. Literature survey

To endeavor with the environment monitoring, momentous accomplishment was born out in shrewd and diminish this technique. The manifestation of environment monitoring and WSN related facts are premeditated by a lot of investigator and have proclaimed demographic data incidents. In 2008, Yang et al. [7] disclosed, "An Environmental Monitoring System with Integrated Wired and Wireless Sensors" fixate a novel environmental monitoring system with a concentrate on the comprehensive system planning for smooth alliance of wired and wireless sensors for long-term, inaccessible monitoring. A consolidated plan for sensor data assembly, execution, promulgation was also presented in this paper.

In 2009, Flammini et al. [8] reported, "Wired and wireless sensor networks for industrial applications" noted a real-time sensor networks for industrial applications. Particular consideration has been compensated to the explanation of arrangements and avenue for completion evaluation was conferred. This paper represents the limitations of wired network and how it is overcome by wireless sensor network.

Kaur et al. [9], in 2014, narrated, "Comparisons of wired and wireless networks: A review", revealed the resemblance between wired and wireless networking on the basis of disparate hardware demand, ranges, flexibility, accuracy and assets. Wired and Wireless networks are more trivial in the private sectors as well as in the household applications. The wired networks administer a defended and swift connectivity but the need of movability, i.e., in any place, anytime is sway the network users close to wireless technology.

3. Wired communication technology

The general block diagram of wired environment monitoring system is shown in **Figure 1**.

This system helps to make the cities pollution free. It monitors the contaminant air and informs about the level of pollution in the air. The wired approach of the system consists of SY-HS-220 humidity sensor, MQ-135 air quality sensor, LDR, LM-35 temperature sensor etc. These sensors output are connected at analog inputs of an Arduino microcontroller. All these sensors are placed over the area to detect the different levels of pollution. The system also makes use of Arduino module, LCD, buzzer and LED's. The LCD screen is used to display the level of pollution within Solapur University campus. It exhibits the category of pollution level. The system puts on buzzer when the level of pollution crosses its threshold limit. Thus this system helps to keep the Solapur University campus pollution free by informing about pollution levels of the areas. This system cost effective and portable. The circuit set up for wired system and its connections are shown in **Figure 2**. The system using wired network is portable and effective. But, drawback of such a system is that cables requirement for providing linkage between the devices. As number of

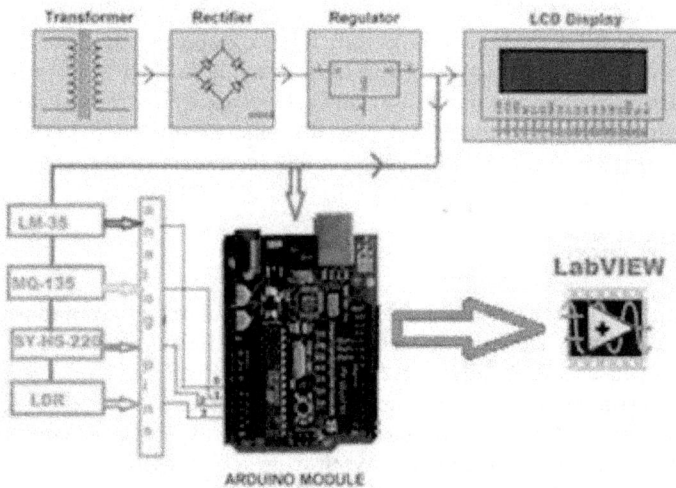

Figure 1.
General block diagram of wired system.

peripheral increased in the system, it leads to lofty installation and protection costs, e.g., due to low scalability and more breakdown rate of connectors. Consequently, wireless technology is the best solution for todays (**Figures 3** and **4**).

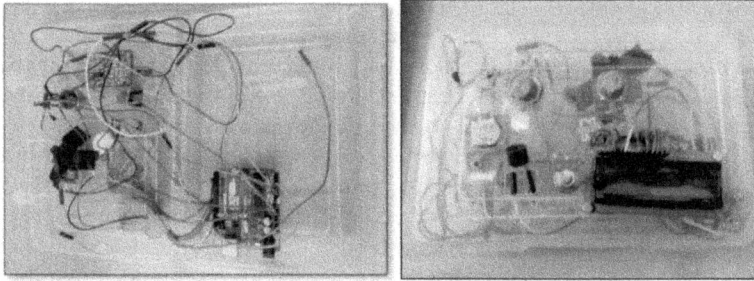

Figure 2.
Circuit connection of a system.

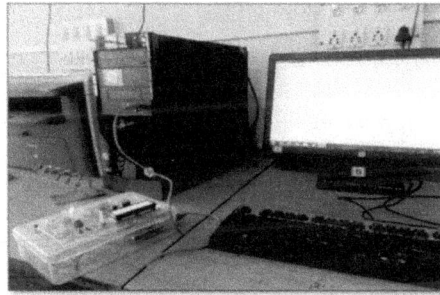

Figure 3.
Experimental set up for wired environment monitoring system.

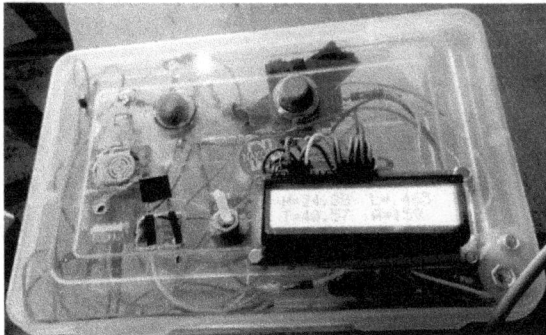

Figure 4.
Actual readings of each sensor on LCD.

4. Wireless communication technology

Due to advancement in technology wireless network being used to avoid cabling cost and to obtain efficient control. We proposed to use WSN technology for it. In today's world, the wireless sensor networks (WSN) is one of the most momentous technology. The monitoring, reorganization and controlling of the data are the key concern of this technology. The inaccessible interface and actual monitoring with the physical world can be done easily by sensor node of the network. The wireless sensor networks differ from general data networks, because WSN are application oriented, planned and deployed for dedicated purpose [10].

The whole system was designed using ATMega 328 microcontroller integrated with XBee S2 protocol to form sensing phenomena. The planning of mote consists of a processing entity conscientious for compilation and giving out the data sensed by a sensor. A radio transceiver mechanism used as a communication part accompanied by the sensors and a battery is the power provide unit in this system. We anticipated four sensors for measuring temperature, humidity, air quality and light intensity within Solapur University campus. The humidity SY-HS 220 sensor module is used for measuring humidity. Its operating voltage and temperature is 5 V, 0–60°C. The −30 to 85°C is storage temperature range of this module. It converts relative humidity to voltage and can be used in environment monitoring applications. LDR sensor and its voltage divider circuit were used to measured light intensity. LM 35 temperature sensor gives 10 mV per 1° rise in temperature. While, MQ-135 performs ambient air quality monitoring. The circuit connection of sensors to Arduino is shown in **Figure 2**. The developed WSN system uses two motes and one sink node.

The general block diagram of wireless environment monitoring system is shown in **Figure 5**.

According to **Figure 5**, working of the proposed system is carried out. Initially, we have calibrated the individual sensors and then connected to analog inputs of an Arduino microcontroller. This controller integrated with 10-bit ADC, which renovate the analog signals into digital output (**Figures 6** and 7).

The results are displayed on the serial monitor window of Arduino. Through VISA function tool, it is displayed on the front panel of the LabVIEW. For this, we used LabVIEW run time engine, which means that without installation of LabVIEW on your computer, you can run any LabVIEW program, which reduces cost of the system. The developed GUI (Graphical User Interface) using LabVIEW for system continuously monitors the environment data [11]. In X-CTU software, individual ID's for each motes are specified [12–14] (**Tables 1** and 2).

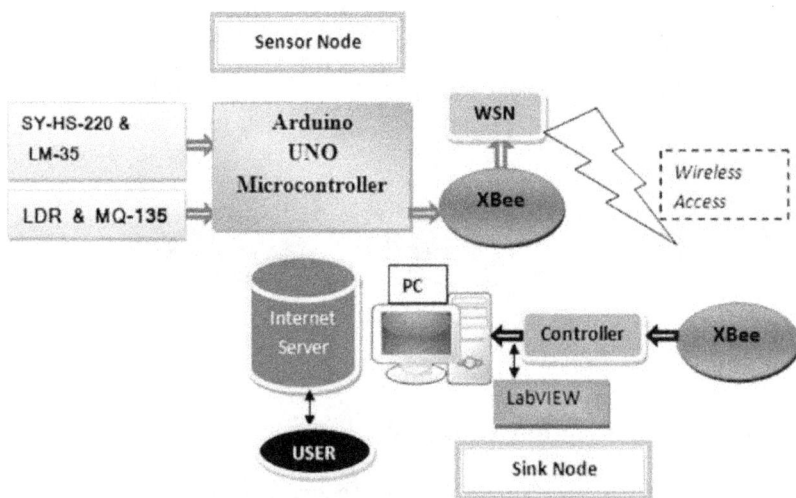

Figure 5.
General block diagram of wireless system.

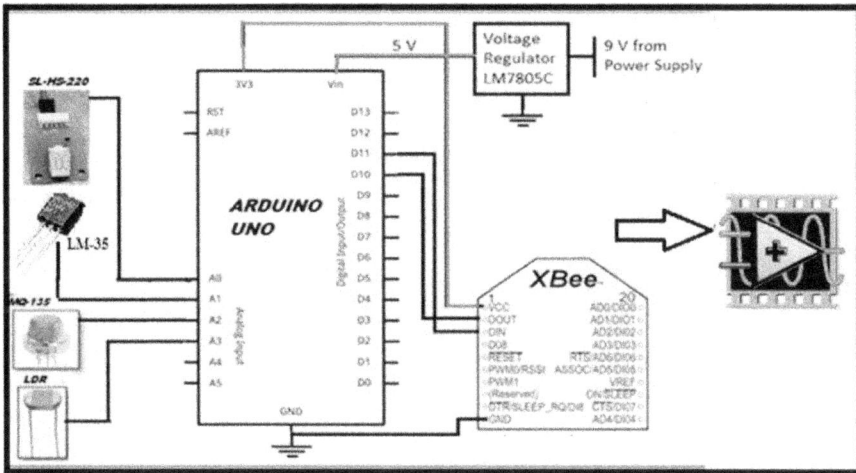

Figure 6.
Circuit connection of sensor node.

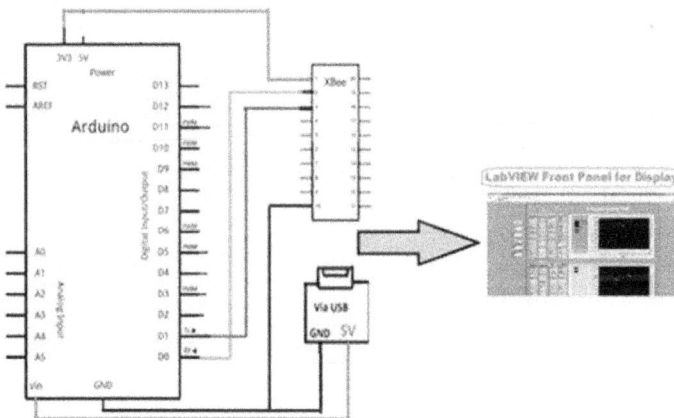

Figure 7.
Circuit connection of sink node.

The data from each mote was separated using LabVIEW software as follows:

- The sink node (Arduino UNO board) was interfaced to LabVIEW through VISA (Virtual Instrument Software Architecture),

- Sink node is a common receiver, which receives data from several motes,

- After this division of the data was carried out.

The LabVIEW GUI was used to monitor the environment quality level. LabVIEW is scheme-intend software that allows to program tools on a GUI for the measurement and control of the systems.

LabVIEW is a graphical improvement tool developed by a National Instruments. These tools are very interactive and superficial for encoding. We can amend the

Sr. no.	Parameters name	Parameters symbol	Configuration value
1.	PAN ID	ID	100
2.	Destination address high	DH	0
3.	Destination address low	DL	0
4.	Scan channel	SC	FFFF(Hex Value)
5.	Scan duration	SD	3
6.	Channel verifications	JV	1
7.	Device option	DO	1
8.	Node identifier	NI	Node 1 to Node 2
9.	Node join time	NJ	FF(Hex Value)
10.	Node discovery back off	NT	3C
11.	Power level	PL	4
12.	Power mode	PM	1
13.	Power at PL4	PP	3
14.	Baud rate	BD	3
15.	RSSI PWM timer	RP	28
16.	DI07 configuration	DI07	1
17.	DI06 configuration	DI06	0
18.	IO sampling rate	IR	3E8
19.	Parity	NB	0
20.	RSSI of last packet	DB	0

Table 1.
XBee parameter for router.

Sr. no.	Parameters name	Parameters symbol	Configuration value
1.	PAN ID	ID	100
2.	Destination address high	DH	0
3.	Destination address low	DL	0
4.	Scan channel	SC	FFFF(Hex Value)
5.	Scan duration	SD	3
6.	Channel verifications	JV	0
7.	Device option	DO	1
8.	Node identifier	NI	—
9.	Node join time	NJ	FF
10.	Node discovery back off	NT	3C
11.	Power level	PL	4
12.	Power mode	PM	1
13.	Power at PL4	PP	3
14.	Baud rate	BD	3
15.	RSSI PWM timer	RP	28
16.	DI07 configuration	DI07	1

Sr. no.	Parameters name	Parameters symbol	Configuration value
17.	DIO6 configuration	DIO6	0
18.	IO sampling rate	IR	3E8
19.	Parity	NB	0
20.	RSSI of last packet	DB	0

Table 2.
XBee parameter for coordinator.

indoctrination gush as we want. The proficient machine code is the distinctive chattels of LabVIEW. The developed G-code of LabVIEW is more indulgent and required execution time is less. The freeware driver makes it more intuitive. The communication, instrumentation, neural networking, control system etc. tools in the LabVIEW have its own task to engender G-code relating to this. The current wireless system uses web publishing tool to show the monitored information on the web page for distant monitoring.

5. Results

When sensor nodes are placed within Solapur University campus, it continuously monitors an environment, the readings from each sensor node will send to the gateway node. This gateway node will send the data to LabVIEW through VISA. The graphical representation of sensor node 1 and sensor node 2 are depicted in the **Figures 8** and **9**.

Figure 10 represents the GUI crated by LabVIEW for environment monitoring system.

The LabVIEW **programming** for this was done as follows (**Figure 11**).

For real time monitoring of environment system, we used a web publishing tool in LabVIEW. This tool was used for web portal connectivity to cover stout monitoring vicinity.

By accessing the web server, inaccessible monitoring and controlling of this system was done using a web publishing tool. Based on the parameters

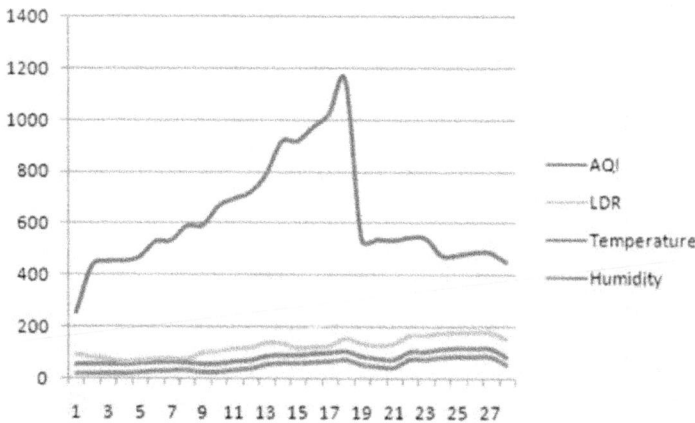

Figure 8.
Graphical representation of sensor node 1.

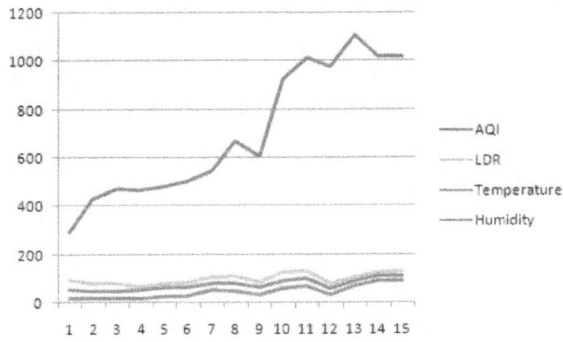

Figure 9.
Graphical representation of sensor node 2.

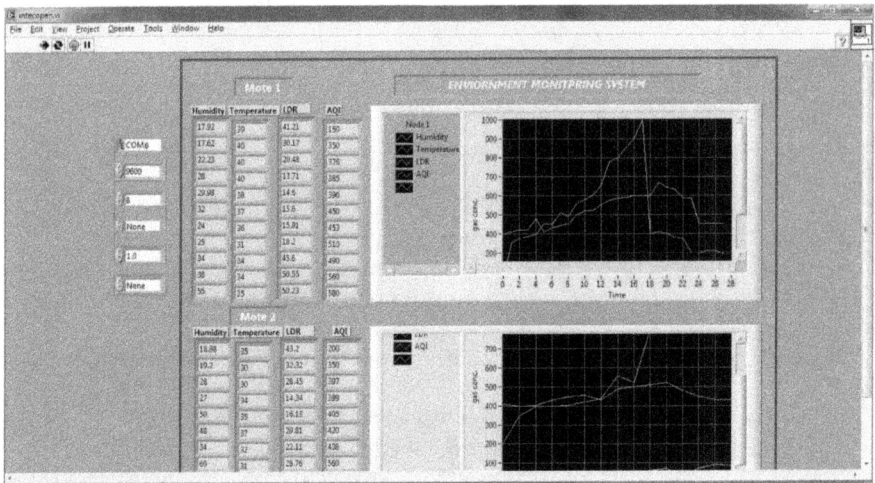

Figure 10.
GUI of monitoring environmental parameters of sensor nodes.

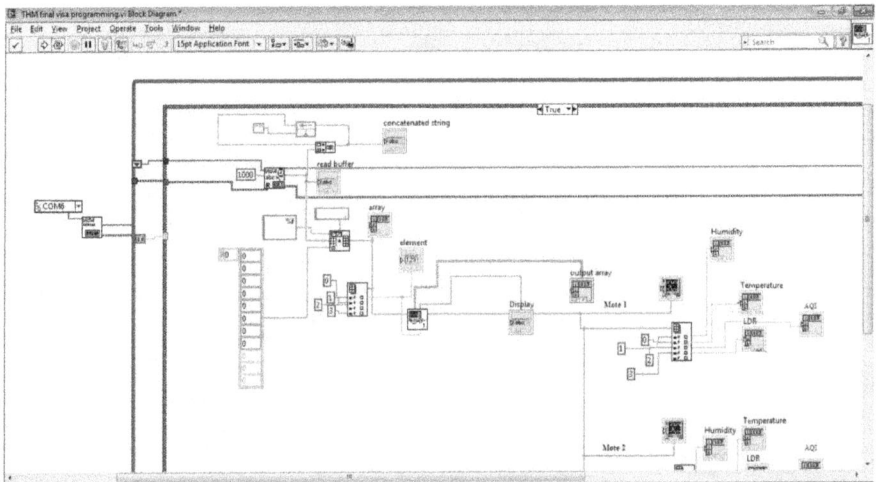

Figure 11.
The G-code for sensor nodes for environment monitoring.

specified in the program, this tool converts the front panel into HTML web page. **Figure 13** represents the GUI of the web server. Concisely, we built an effortless VI that monitors the wireless system. This application was launched on the internet and monitored it tenuously and controlled it involuntarily. For initiation this application on the internet, we must have to arrange the web access. Port address of LabVIEW is 8000. By enabling the various setting in a web publishing tool, we have a right of entry to access other unapproachable panel server and all other IP addresses which we desire. The URL obtained from the LabVIEW page is http://dell-pc:8000/intecopen.html. **Figure 12** shows the data monitoring system in the internet browser before putting a control over VI. When we put a control over VI, it is shown in **Figure 13**. The internet browsing of a system helps to monitor an environment quality at remote places continuously to the users. It is shown in **Figure 14**.

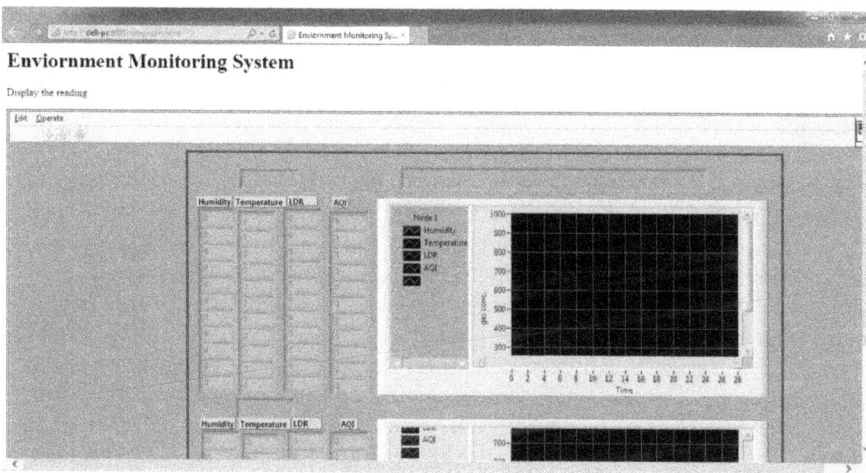

Figure 12.
The GUI of a system displayed on the web server before putting control over VI.

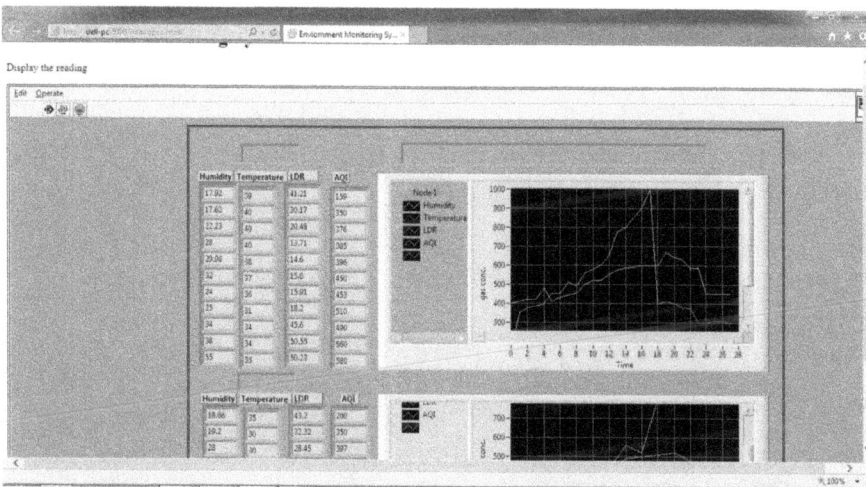

Figure 13.
Distant monitoring of a system after putting control over VI.

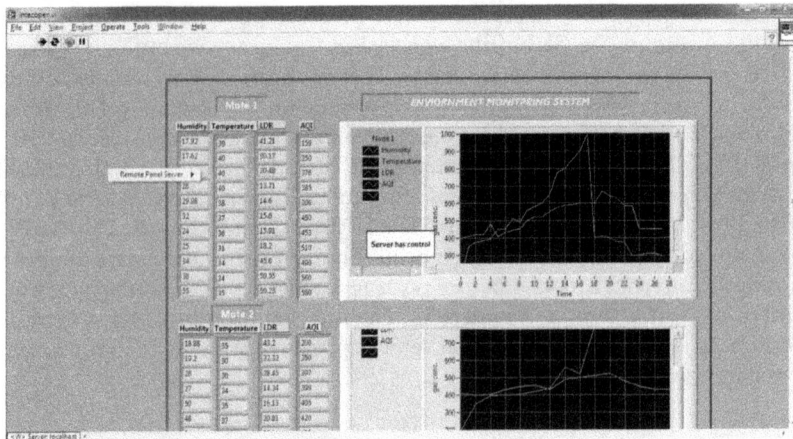

Figure 14.
Front panel showing remote monitoring of a system.

6. Conclusions

A lucrative environmental monitoring method with least amount of components has been constructed. The system is successfully developed using wired and wireless networks. The limitations of wired network and opportunities using wireless networks are rigorously described. The environment monitoring sensors with an Atmega 368 microcontroller, Web portal is proposed. For sending and receiving of the data, the web publishing tool in LabVIEW is used. The system is developed using two motes and one sink node. XBee protocol is used to provide wireless access. This system provides a real-time monitoring via money-spinning low data rate and significant power wireless communication technology. We envisage that this system will encompass an enormous recognition in the industrialized sectors and will realize an effective amalgamation among WSN and Web portal. Accordingly, a tack target of inaccessible monitoring of the air quality within the environment can be attained. It is highly pertinent to metrological departments and also in industrial sectors. In future, we would be fond of to be made controlling system for environment monitoring.

Author details

Tabbsum Mujawar* and Lalasaheb Deshmukh
School of Physical Sciences, Solapur University, Solapur, Maharashtra, India

*Address all correspondence to: thmujawar@sus.ac.in

IntechOpen

References

[1] Mujawar T. Development of wireless sensor network for hazardous gas detection and alert system [thesis]. Inflibnet. 30 Jan 2017. Available from: http://www.shodhgang.inflibnet.ac.in

[2] Wang N, Zhang N, Wang M. Wireless sensors in agriculture and food industry. Recent development and future perspective. Computers and Electronics in Agriculture. 2006;**50**:1-14

[3] Suryadevara N, Mukhopadhyay S. Wireless sensor network based home monitoring system for wellness determination of elderly. IEEE Sensors Journal. 2012;**12**:1965-1972

[4] Tiantian J, Zhanyong Y. Research on mine safety monitoring system based on WSN. Procedia Engineering. 2011;**26**:2146-2151

[5] Tilak S, Ghazaleh N, Heinzelman W. A taxonomy of wireless micro sensor network nodels. ACM SIGMOBILE Mobile Computing and Communications Review. 2002;**6**:28-36

[6] Margolis M. Arduino Cookbook. 1st ed. Massachusetts, United States: O'REILLY® Media, Inc.; 2011. pp. 81-213

[7] Yang J, Zhang C, Li X, Huang Y, Fu S, Acevedo M. An environmental monitoring system with integrated wired and wireless sensors. In: International Conference on Wireless Algorithms, Systems and Applications; 2008;**3**:224-236

[8] Flammin A, Ferrari P, Marioli D, Sisinni E. Wired and wireless sensor networks for industrial applications. Microelectronics Journal. 2009;**40**:1322-1336

[9] Kaur N, Monga S. Comparisons of wired and wireless networks: A review. International Journal of Advanced Engineering Technology. 2014;**2**:34-35

[10] Mujawar T, Kasbe M, Mule S, Deshmukh L. Development of wireless gas sensing system for home safety. International Journal of Engineering Sciences & Emerging Technologies. 2016;**8**:213-221

[11] Mujawar T, Kasbe M, Mule S, Deshmukh L. Online monitoring of WSN based air quality monitoring system. AIP Conference Proceedings. 2018;**02003**:020030-1-020030-9

[12] Schell M, Guvench M. Development of a general purpose XBee series-2 API-mode communication library for LabVIEW. In: Northeast Section Conference (ASEE); April 27-28, 2012

[13] Zhang J, Song G. Design of a wireless sensor network based monitoring system for home automation. In: International Conference on Future Computer Sciences and Application. June 2011;7:18-19

[14] Li Y, Ji M. Design of home automation system based on ZigBee wireless sensor network. In: 1st International Conference on Information Science and Engineering (ICISE); December 26-28, 2009

www.ingramcontent.com/pod-product-compliance
Lightning Source LLC
Chambersburg PA
CBHW081234190326
41458CB00016B/5781